INVITATION TO NUMBER THEORY

NEW MATHEMATICAL LIBRARY

PUBLISHED BY

THE MATHEMATICAL ASSOCIATION OF AMERICA

Editorial Committee

Basil Gordon, Chairman (1975-76)   Anneli Lax, Editor
University of California, L.A.   *New York University*

Ivan Niven (1975-77)   *University of Oregon*
M. M. Schiffer (1975-77)   *Stanford University*

The New Mathematical Library (NML) was begun in 1961 by the School Mathematics Study Group to make available to high school students short expository books on various topics not usually covered in the high school syllabus. In a decade the NML matured into a steadily growing series of some twenty titles of interest not only to the originally intended audience, but to college students and teachers at all levels. Previously published by Random House and L. W. Singer, the NML became a publication series of the Mathematical Association of America (MAA) in 1975. Under the auspices of the MAA the NML will continue to grow and will remain dedicated to its original and expanded purposes.

INVITATION TO NUMBER THEORY

by

Oystein Ore

Yale University

20

MATHEMATICAL ASSOCIATION
OF AMERICA

Acknowledgments

Benjamin Franklin's magic squares and circles are published through the courtesy of the Yale Franklin Collection.

The Adam Riese picture is published through the courtesy of the Beinecke Library at Yale.

Illustrated by George H. Buehler

Seventh Printing

© Copyright 1967, by the Mathematical Association of America
All rights reserved under International and Pan-American Copyright
Conventions. Published in Washington by
The Mathematical Association of America

Library of Congress Catalog Card Number: 67-20607

Complete Set ISBN-0-88385-600-X

Vol. 20 0-88385-620-4

Manufactured in the United States of America

NEW MATHEMATICAL LIBRARY

MAA Service Center
P. O. Box 91112
Washington, DC 20090-1112
800-331-1622   fax: 301-206-9789

Contents

INVITATION TO NUMBER THEORY

CHAPTER ONE

Introduction

1.1 History

Number theory is a branch of mathematics which deals with the *natural numbers*,

$$1, \ 2, \ 3, \cdots,$$

often called the *positive integers*.

Archeology and history teach us that man began early to count. He learned to add numbers and much later to multiply and subtract them. To divide numbers was necessary in order to share evenly a heap of apples or a catch of fish. These operations on numbers are called calculations. The word "calculation" is derived from the Latin *calculus*, meaning a little stone; the Romans used pebbles to mark numbers on their computing boards.

As soon as men knew how to calculate a little, this became a playful pastime for many a speculative mind. Experiences with numbers accumulated over the centuries with compound interest, so to speak, till we now have an imposing structure in modern mathematics known as number theory. Some parts of it still consist of simple play with numbers, but other parts belong to the most difficult and intricate chapters of mathematics.

1.2 Numerology

Some of the earliest traces of number speculations can certainly be detected in superstitions concerning numbers, and these one finds among all peoples. There are lucky numbers to be preferred and cherished, and there are unlucky ones to be shunned like the evil eye. We have a good deal of information about the *numerology* of the classical Greeks, that is, their thoughts and superstitions in regard to the symbolic meaning of the various numbers. For instance, an odd number greater than one symbolized a male idea, and even numbers represented females; hence the number 5, the sum of the first male and female number, symbolized marriage or union.

Any one who wishes examples of more advanced numerology may take Plato's *Republic* out of the library and read Book 8. While such numerology does not represent much in the way of mathematical ideas, it does involve manipulations of numbers and their properties. And as we shall see a little later, some remarkable problems in number theory still occupying mathematicians have their origins in Greek numerology.

As regards number superstitions, we have presently no great cause for feeling superior. We all know hostesses who for no price would have 13 guests at the table, and there are remarkably few hotels having room number or floor number 13. We really don't know why such number taboos arise. There are many plausible explanations, but most of them without much foundation; for example, we are reminded that there were 13 guests at the Last Supper, the 13th of course being Judas. The observation that many things are counted in dozens, and 13 gives us a "baker's dozen" with an odd item left over, may be more realistic.

In the Bible, particularly in the Old Testament, the number 7 plays a special role; in old Germanic folklore the numbers 3 and 9 are often repeated; and the Hindus in their mythology were very partial to the number 10.

1.3 The Pythagorean Problem

As an example of early number theory we may mention the *Pythagorean problem*. As we know, in a right-angled triangle the lengths of the sides satisfy the Pythagorean relation

$$(1.3.1) \qquad z^2 = x^2 + y^2,$$

where z is the length of the hypotenuse. This makes it possible in a right triangle to compute the length of one side when one knows the

other two. Incidentally, naming this theorem for the Greek philos-
opher Pythagoras is somewhat inappropriate; it was known to the
Babylonians nearly 2000 years before his time.

Sometimes all the side lengths x, y, z in (1.3.1) are integers. The
simplest case,

(1.3.2) $x = 3$, $y = 4$, $z = 5$,

has been found on Babylonian tablets. One can interpret it as follows.
Suppose we have a rope hoop with marks or knots at equal intervals
dividing it into 12 parts; then when we stretch the hoop on three
pegs in the field so that we obtain a triangle with sides 3 and 4, the
third side has length 5 and its opposite angle is right (Figure 1.3.1).
Often one reads in histories of mathematics that this was the method
of constructing a right angle used by the Egyptian surveyors or
harpedonapts (rope stretchers) in laying out fields after the inunda-
tions of the Nile. However, this may well be one of the many myths
in the history of science; we have no contemporary evidence sup-
porting it.

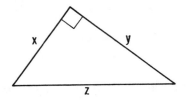

Figure 1.3.1

There are many other cases of integral solutions of the Pythagorean
equation (1.3.1), for example,

$$x = 5, \quad y = 12, \quad z = 13,$$

$$x = 7, \quad y = 24, \quad z = 25,$$

$$x = 8, \quad y = 15, \quad z = 17.$$

We shall show later how all such solutions can be found. The Greeks
knew how to determine them, and probably the Babylonians did also.

When two integers x and y are given, one can always find a
corresponding z satisfying (1.3.1), but z may well be irrational.
When one requires all three numbers to be integral, the possibilities
are severely limited. The Greek mathematician Diophantos of

Alexandria (date uncertain, about 200 A.D.) wrote a book *Arith-metica* which deals with such problems. Since his time the question of finding integral or rational solutions of equations is called a Diophantine problem, and Diophantine analysis is an important part of present day number theory.

Problem Set 1.3

1. Try to find other solutions in integers of the Pythagorean equation.

2. Try to find other solutions where the hypotenuse is one unit larger than the larger of the two legs.

1.4 Figurate Numbers

In number theory we often encounter square numbers like

$$3^2 = 9, \qquad 7^2 = 49, \qquad 10^2 = 100$$

and similarly cube numbers such as

$$2^3 = 8, \qquad 3^3 = 27, \qquad 5^3 = 125.$$

This geometric manner of expression is one of our many legacies from Greek mathematical thought. The Greeks preferred to think of numbers, including the integers, as geometric quantities. Consequently, a product $c = a \cdot b$ was thought of as the area c of a rectangle with sides a and b. One could also think of $a \cdot b$ as the number of dots in a rectangular array with a dots on one side and b dots on the other. For instance, $20 = 4 \cdot 5$ is the number of dots in the rectangular array of Figure 1.4.1.

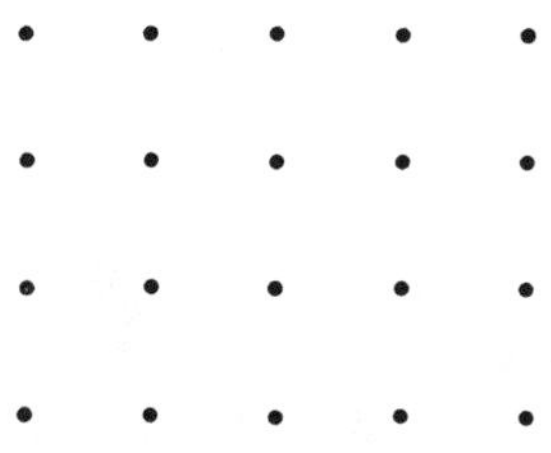

.   .   .   .   .

.   .   .   .   .

.   .   .   .   .

.   .   .   .   .

Figure 1.4.1

Any integer which is a product of two integers could be called a *rectangular number*. When the two sides of the rectangle have the same length, the number is a square number. Some numbers cannot be represented as rectangular numbers except in the trivial way that one strings along the points in a single row; for instance 5 can be represented as a rectangular number only by taking one side to be 1 and the other to be 5 (Figure 1.4.2). Such numbers the Greeks called *prime numbers*. A single point would usually not be considered a number. The unit 1 was the brick from which all the proper num bers were built up. *Thus 1 was not, and is not, a prime number*

Figure 1.4.2

Instead of rectangles and squares one could consider points located regularly within other geometric figures. In Figure 1.4.3 we have illustrated successive *triangular numbers*.

In general, the nth triangular number is given by the formula

$$(1.4.1) \qquad T_n = \tfrac{1}{2}n(n+1), \qquad n = 1, 2, 3, \cdots.$$

These numbers have a variety of properties; for instance, the sum of two consecutive triangular numbers is a square:

$$(1.4.2) \quad 1 + 3 = 4, \qquad 3 + 6 = 9, \qquad 6 + 10 = 16, \qquad \text{etc.}$$

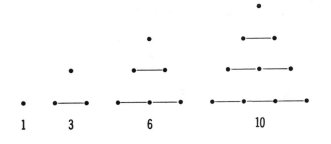

Figure 1.4.3

The triangular and square numbers were generalized to higher polygonal numbers. Let us illustrate this with the pentagonal numbers defined through Figure 1.4.4.

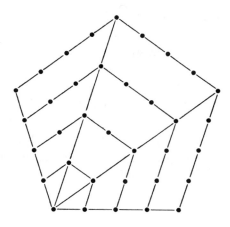

Figure 1.4.4

One reads off that the first few pentagonal numbers are

(1.4.3)   1, 5, 12, 22, 35.

It can be shown that the nth pentagonal number p_n is given by

(1.4.4) $p_n = \frac{1}{2}(3n^2 - n).$

Hexagonal numbers, and in general k-gonal numbers defined by a regular polygon with k sides, are obtained analogously. We shall not spend more time discussing them. The figurate numbers, particularly the triangular numbers, were popular in number studies in the late Renaissance after Greek number theory had been brought to Western Europe; they still occasionally appear in papers on number theory.

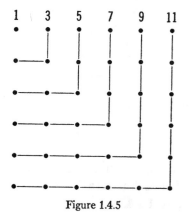

Figure 1.4.5

Several simple number relations can be deduced from such a geometric analysis. Let us point out only one fact. It was early discovered that if one sums the odd numbers up to a certain point, the result is always a square; for example,

$$1 + 3 = 4, \quad 1 + 3 + 5 = 9, \quad 1 + 3 + 5 + 7 = 16, \quad \text{etc.}$$

To prove such a relation one need only glance at the diagram of nested squares which we have drawn in Figure 1.4.5.

Problem Set 1.4

1. Prove the general formula (1.4.1) for triangular numbers by induction.

2. Prove the formula (1.4.4) for the pentagonal numbers.

3. Show that the general expression for a k-gonal number is

$$\tfrac{1}{2}k(n^2 - n) - n^2 + 2n.$$

1.5 Magic Squares

If you have played shuffle board you may recall that the nine squares on which you tried to place your disks were numbered from 1 to 9 and arranged in the following pattern:

2	9	4
7	5	3
6	1	8

Figure 1.5.1

Here the numbers in every row, in every column, and in each of the two diagonals add up to the same total, 15.

In general, a *magic square* is an arrangement of the integers from 1 to n^2 in a square scheme such that the numbers in each row, column, and diagonal give the same sum s, the *magic sum*. As a magic square on $4^2 = 16$ numbers we may take Figure 1.5.2. Here the magic sum is 34.

1	8	15	10
12	13	6	3
14	11	4	5
7	2	9	16

Figure 1.5.2

For each n there is only one magic sum s, and it is easy to see what it must be: The sum of the numbers in each column is s; since there are n columns the sum of all numbers in the magic square is ns. But the sum of all numbers from 1 to n^2 is

$$1 + 2 + \cdots + n^2 = \tfrac{1}{2}(n^2 + 1)n^2,$$

as one sees from the expression for the sum of the numbers in an arithmetic progression. Since

$$ns = \tfrac{1}{2}(n^2 + 1)n^2,$$

it follows that

(1.5.1) $$s = \tfrac{1}{2}n(n^2 + 1);$$

so if n is given, s is determined. Magic squares can be constructed for all n greater than 2; but the reader can easily verify that there is none for $n = 2$.

The strange properties of these squares were considered magical in medieval days and so the squares served as talismans, protecting the wearer against many evils. An often reproduced magic square is the one in Albrecht Dürer's famous print Melancholia (see frontispiece). It is reproduced in greater detail in Figure 1.5.3; incidentally, it shows how the digits were written at the time of Dürer. The middle numbers in the last line represent the year 1514, in which we know that Dürer's print was made. He probably started from these two numbers and found the remaining ones by trial and error.

Figure 1.5.3

We can prove that when $n = 3$ there is essentially only one magic square, namely the one in Figure 1.5.1. To do it we write such a table in the general form

$$
\begin{array}{ccc}
x_1 & y_1 & z_1 \\[4pt]
x_2 & y_2 & z_2 \\[4pt]
x_3 & y_3 & z_3
\end{array}
$$

and examine what these nine numbers can be.

First we show that the central number y_2 must be 5. To verify this we note, using (1.5.1), that for $n = 3$ the magic sum is $s = 15$. Then we sum the terms in the second row, the second column, and the two diagonals. This, as one sees, gives each term once except y_2, which appears four times, since it occurs in each of the four sums.

Therefore, since each sum is s, we have

$$4s \; = \; 4 \times 15 \; = \; 60$$
$$= \; x_2 + y_2 + z_2 + y_1 + y_2 + y_3 + x_1 + y_2 + z_3 + z_1 + y_2 + x_3$$
$$= \; 3y_2 + x_1 + x_2 + x_3 + y_1 + y_2 + y_3 + z_1 + z_2 + z_3$$
$$= \; 3y_2 + 1 + 2 + \cdots + 9$$
$$= \; 3y_2 + 45;$$

hence

$$3y_2 \; = \; 60 - 45 \; = \; 15 \quad \text{and} \quad y_2 \; = \; 5.$$

In the scheme

$$
\begin{array}{ccc}
x_1 & y_1 & z_1 \\
x_2 & 5 & z_2 \\
x_3 & y_3 & z_3
\end{array}
$$

the number 9 cannot occur in a corner; for, if for example $x_1 = 9$, then $z_3 = 1$ (since $s = 15$) and the square would have the form

$$
\begin{array}{ccc}
9 & y_1 & z_1 \\
x_2 & 5 & z_2 \\
x_3 & y_3 & 1
\end{array}
$$

The four numbers y_1, z_1, x_2, x_3 must all be less than 6, since $y_1 + z_1 = x_2 + x_3 = 6$. But we have left only three numbers less than 6, namely 2, 3, and 4, so this is impossible. We conclude that 9 must be placed in the middle of a row or column so that our square may be taken as

$$
\begin{array}{ccc}
x_1 & 9 & z_1 \\
x_2 & 5 & z_2 \\
x_3 & 1 & z_3
\end{array}
$$

The number 7 cannot be in the same row as 9 since the sum would exceed 15; nor can 7 be in the same row as 1, for then the remaining number would also have to be 7. Thus 7 cannot lie in a corner and we can assume that the square has the form

$$x_1 \quad 9 \quad z_1$$

$$7 \quad 5 \quad 3$$

$$x_3 \quad 1 \quad z_3$$

The numbers in the row with 9 must be 2 and 4 since otherwise the sum would exceed 15. Furthermore, the 2 must be in the same column as 7, for if the 4 were there, the third number in this column would also be 4. By this observation the place of the remaining two numbers 6 and 8 are determined and we obtain the magic square shown in Figure 1.5.1

52	61	4	13	20	29	36	45
14	3	62	51	46	35	30	19
53	60	5	12	21	28	37	44
11	6	59	54	43	38	27	22
55	58	7	10	23	26	39	42
9	8	57	56	41	40	25	24
50	63	2	15	18	31	34	47
16	1	64	49	48	33	32	17

Figure 1.5.4

For larger n one can construct a great variety of magic squares; in the sixteenth and seventeenth century and even later magic square making flourished much like crossword puzzles of today. Benjamin Franklin was an eager magic square fan. He confessed later that while he was clerk of the Pennsylvania Assembly and needed to while away the tedium of formal business, he filled out some peculiar magic squares and even magic circles consisting of intertwined circles of numbers whose sums on each circle were the same. The following account is taken from *The Papers of Benjamin Franklin*, Yale University Press, vol. 4, pp. 392–403.

Franklin's magic squares came to light when one of his friends, Mr. Logan, showed him various books on the subject and observed that he did not believe that any Englishman had done anything remarkable of this kind. "He then shewed me several in the same book of an uncommon and more curious kind, but as I thought none of them equal to some I remembered to have made, he desired me to let him see them; and accordingly, the next time I visited him I carried him a square of 8, which I found among my old papers and which I will now give you, with an account of its properties." (Figure 1.5.4.)

Franklin mentions only some of the properties of his square; we leave it to the reader to discover more. One sees that the sum is $s = 260$, but each half row, and half column, adds up to 130 which is half of 260. The four corner numbers with the middle numbers make 260; the bent row from 16 up to 10 and then descending from 23 to 17 makes 260, and so does every parallel bent row of 8 numbers.

"Mr. Logan then shewed me an old arithmetical book, in quarto, wrote, I think, by one Stiefelius [Michael Stiefel, *Arithmetica integra*, Nürnberg, 1544] which contained a square of 16, that he said he should imagine must have been a work of great labour; but if I forget not, it had only the common property of making the same sum, viz., 2056, in every row, horizontal, vertical and diagonal.

"Not willing to be out-done by Mr. Stiefelius, even in the size of my square, I went home, and made, that evening, the following magical square of 16, which, besides having all the properties of the foregoing square of 8, i.e., it would make the sum 2056 in all the same rows and diagonals, had this added that a four square hole being cut in a piece of paper of such a size as to take in and shew through it, just 16 of the little squares when laid on the greater square, the sum of the 16 numbers so appearing through the hole, wherever it was placed on the greater square, should likewise make 2056."

Here is Franklin's magic square and you may yourself try out its remarkable properties. (Figure 1.5.5.)

Franklin was justifiably proud of his creation as one may see from the sequel of his letter: "This I sent to our friend the next morning, who, after some days, sent it back in a letter, with these words: 'I return to thee thy astonishing or most stupendous piece of the magical square in which . . . ,' but the compliment is too extravagant, and therefore, for his sake, as well as my own, I ought not to repeat it. Nor is it necessary, for I make no question but you will readily allow this square of 16 to be the most magically magical of any magic square ever made by any magician."

For more information on the construction of magic squares see J. V. Uspensky and M. A. Heaslet, *Elementary Number Theory*, McGraw-Hill, New York, 1939.

A Magic Square of Squares.

200	217	232	249	8	25	40	57	72	89	104	121	136	153	168	185
58	39	26	7	250	231	218	199	186	167	154	135	122	103	90	71
198	219	230	251	6	27	38	59	70	91	102	123	134	155	166	187
60	37	28	5	252	229	220	197	188	165	156	133	124	101	92	69
201	216	233	248	9	24	41	56	73	88	105	120	137	152	169	184
55	42	23	10	247	234	215	202	183	170	151	138	119	106	87	74
203	214	235	246	11	22	43	54	75	86	107	118	139	150	171	182
53	44	21	12	245	236	213	204	181	172	149	140	117	108	85	76
205	212	237	244	13	20	45	52	77	84	109	116	141	148	173	180
51	46	19	14	243	238	211	206	179	174	147	142	115	110	83	78
207	210	239	242	15	18	47	50	79	82	111	114	143	146	175	178
49	48	17	16	241	240	209	208	177	176	145	144	113	112	81	80
196	221	228	253	4	29	36	61	68	93	100	125	132	157	164	189
62	35	30	3	254	227	222	195	190	163	158	131	126	99	94	67
194	223	226	255	2	31	34	63	66	95	98	127	130	159	162	191
64	33	32	1	256	225	224	193	192	161	160	129	128	97	96	65

B.Franklin inv. I.Ferguson delin.   J.Mynde fc.

Figure 1.5.5

Problem Set 1.5

1. When Dürer constructed his magic square (Figure 1.5.3) could he have used other squares with the year marked in the same way?

2. Dürer lived till the year 1528. Could he have dated any of his later pictures in the same manner?

3. Study some of the properties of Franklin's magic circles (page 14).

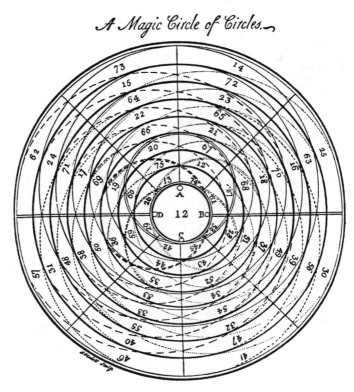

Figure 1.5.6. This is a reproduction of Franklin's magic circles. The original which is in colors was recently sold to a private collector at an auction in New York.

CHAPTER TWO

Primes

2.1 Primes and Composite Numbers

It must have been one of the first properties discovered that some numbers can be factored into two or more smaller factors, for example,

$$6 = 2 \cdot 3, \quad 9 = 3 \cdot 3, \quad 30 = 2 \cdot 15 = 3 \cdot 10,$$

while others like

$$3, \quad 7, \quad 13, \quad 37$$

cannot be so factored. Let us recall that in general when

$$(2.1.1) \qquad\qquad c = a \cdot b$$

is a product of two numbers a and b, we call a and b *factors* or *divisors* of c. Every number has the *trivial factorization*

$$(2.1.2) \qquad\qquad c = 1 \cdot c = c \cdot 1.$$

Correspondingly we call 1 and c *trivial divisors* of c.

Any number $c > 1$ which has some non-trivial factorization we call *composite*. When c has only the trivial factorization (2.1.2), it is called a *prime*. Among the first 100 numbers the following 25 are primes:

$$2, \quad 3, \quad 5, \quad 7, \quad 11, \quad 13, \quad 17, \quad 19, \quad 23, \quad 29, \quad 31, \quad 37, \quad 41,$$

$$43, \quad 47, \quad 53, \quad 59, \quad 61, \quad 67, \quad 71, \quad 73, \quad 79, \quad 83, \quad 89, \quad 97.$$

All the remaining ones, except 1, are composite. We notice:

15

THEOREM 2.1.1. *Any integer $c > 1$ is either a prime or has a prime factor.*

PROOF. If c is not a prime it has a smallest non-trivial factor p. Then p is a prime, for if p were composite, c would have a still smaller factor.

We have now been brought to our first important problem in number theory: How can one decide whether a number is a prime or not; and in case it is composite how can one find a non-trivial divisor?

An immediate but very unsatisfactory answer is that we could try to divide the given number c by all numbers less than it. According to Theorem 2.1.1 it suffices to divide by all primes less than c. But we can reduce this task materially by the remark that in a factorization (2.1.1) not both factors a and b can be greater than $\sqrt{c}$. If this were the case one would have

$$a \cdot b > \sqrt{c} \cdot \sqrt{c} = c$$

which is impossible. Thus to find whether c has a divisor we need only examine whether any of the primes less than or equal to $\sqrt{c}$ divides c.

Example 1. If $c = 91$ then $\sqrt{c} = 9. \cdots$; by trying the primes 2, 3, 5, 7 one sees that $91 = 7 \cdot 13$.

Example 2. If $c = 1973$ one finds $\sqrt{c} = 44. \cdots$. Since none of the primes up to 43 divides c, this number is prime.

One realizes quickly that for large numbers this method may be very cumbersome. However, here as in many other calculations of number theory, one can rely on modern techniques. It is simple to program a computer to divide a given number c by all integers up to $\sqrt{c}$ and print those which give no remainder, that is, those which divide c.

Another very simple method is to rely upon tables of primes, that is, make use of examinations of primes already made by others. Through the last couple of centuries many prime tables have been computed and printed. The most extensive available one is the table by D. N. Lehmer giving all primes up to 10,000,000. Our Table 1 contains all primes to 1000.

Some enthusiastic calculators have actually prepared prime tables for numbers beyond 10,000,000. However, there does not seem to be much point in going to the considerable expense and effort to have them printed. It is only rarely that a mathematician, even a specialist in number theory, runs into the question of deciding whether or not a very large number is a prime. Furthermore, a mathematician does not arbitrarily pounce upon a large number to find out whether it is composite or prime. The numbers he wants to examine usually appear in specific mathematical problems and so have very special forms.

TABLE 1. PRIMES AMONG THE FIRST THOUSAND NUMBERS.

2, 3, 5, 7, 11, 13, 17, 19, 23, 29, 31, 37, 41, 43, 47, 53, 59, 61, 67, 71, 73, 79, 83, 89, 97,

101, 103, 107, 109, 113, 127, 131, 137, 139, 149, 151, 157, 163, 167, 173, 179, 181, 191, 193, 197, 199,

211, 223, 227, 229, 233, 239, 241, 251, 257, 263, 269, 271, 277, 281, 283, 293,

307, 311, 313, 317, 331, 337, 347, 349, 353, 359, 367, 373, 379, 383, 389, 397,

401, 409, 419, 421, 431, 433, 439, 443, 449, 457, 461, 463, 467, 479, 487, 491, 499,

503, 509, 521, 523, 541, 547, 557, 563, 569, 571, 577, 587, 593, 599,

601, 607, 613, 617, 619, 631, 641, 643, 647, 653, 659, 661, 673, 677, 683, 691,

701, 709, 719, 727, 733, 739, 743, 751, 757, 761, 769, 773, 787, 797,

809, 811, 821, 823, 827, 829, 839, 853, 857, 859, 863, 877, 881, 883, 887,

907, 911, 919, 929, 937, 941, 947, 953, 967, 971, 977, 983, 991, 997.

Problem Set 2.1

1. Which of the following numbers are primes:

    (a) The year of your birth?
    (b) The present year number?
    (c) Your house number?

2. Find the next prime larger than the prime 1973.

3. Note that 90 to 96 inclusive are seven consecutive composite numbers; find nine consecutive composite numbers.

2.2 Mersenne Primes

A prime race has been going on for several centuries. Many mathematicians have vied for the honor of having discovered the greatest known prime. One could of course select some very large numbers with no obvious divisors like 2, 3, 5, 7 and try out whether they might be primes. This, as one soon discovers, is not a very effective way, and the race has now settled down to follow a single track that has proved successful.

The Mersenne primes are the primes of the special form

(2.2.1) $M_p = 2^p - 1,$

where p is another prime. These numbers came into mathematics early and appear in Euclid's discussion of the perfect numbers, which we shall encounter later on. They are named for the French friar Marin Mersenne (1588–1648) who calculated a good deal on the perfect numbers.

When one starts calculating the numbers (2.2.1) for various primes p one sees that they are not all primes. For example,

$$M_2 = 2^2 - 1 = 3 = \text{prime},$$
$$M_3 = 2^3 - 1 = 7 = \text{prime},$$
$$M_5 = 2^5 - 1 = 31 = \text{prime},$$
$$M_7 = 2^7 - 1 = 127 = \text{prime},$$
$$M_{11} = 2^{11} - 1 = 2047 = 23 \cdot 89.$$

The general program for finding large primes of the Mersenne type is to examine all the numbers M_p for the various primes p. The numbers increase very rapidly and so do the labors involved. The reason why the work is manageable even for quite large numbers is that there are very effective ways for finding out whether these special numbers are primes.

There was an early phase in the examination of Mersenne primes which culminated in 1750 when the Swiss mathematician Euler established that M_{31} is a prime. By that time eight Mersenne primes had been found, corresponding to the values

$$p = 2, \quad p = 3, \quad p = 5, \quad p = 7,$$
$$p = 13, \quad p = 17, \quad p = 19, \quad p = 31.$$

Euler's number M_{31} remained the greatest known prime for more than a century. In 1876 the French mathematician Lucas established that the huge number

$M_{127} \;=\; 170\,141\,183\,460\,469\,231\,731\,687\,303\,715\,884\,105\,727$

is a prime. It is quite a number, having 39 digits! The Mersenne primes less than it are given by the values of p above and, in addition, by

$$p \;=\; 61, \qquad p \;=\; 89, \qquad p \;=\; 107.$$

These 12 Mersenne primes were calculated by means of just pen on paper and, for some of the later ones, mechanical desk calculators. The introduction of motor powered electric calculators made it possible to continue the search up to $p \;=\; 257$, but the results were disappointing; no further Mersenne primes were found.

This was the situation when computers took over. With the development of larger capacity machines it was possible to push the search for Mersenne primes to higher and higher limits. D. H. Lehmer established that the values

$$p \;=\; 521, \quad p \;=\; 607, \quad p \;=\; 1279, \quad p \;=\; 2203, \quad p \;=\; 2281$$

yield Mersenne primes M_p. Lately there has been further progress. Riesel (1958) showed that

$$p \;=\; 3217$$

yields a Mersenne prime, and Hurwitz (1962) found the two values

$$p \;=\; 4253, \qquad p \;=\; 4423.$$

A huge advance was made by Gillies (1964) who found Mersenne primes corresponding to

$$p \;=\; 9689, \qquad p \;=\; 9941, \qquad p \;=\; 11213.$$

This gives a total harvest so far of 23 Mersenne primes, and as the machine capacities increase we may hope for further ones. Lucas's prime M_{127}, as we mentioned, has 39 digits. Even to compute the largest known prime M_{11213} is quite a task, and there does not seem to be much point in reproducing it here. But we might be interested in finding out how many digits it contains. This we can do as follows, without actually computing the number.

Instead of finding the number of digits in $M_p = 2^p - 1$, let us take the next number,

$$M_p + 1 = 2^p.$$

These two numbers must have the same number of digits for, if $M_p + 1$ should have one more digit, it would have to be a number which ends in 0. But this is not possible for any power of 2, as one sees from the series

$$2, 4, 8, 16, 32, 64, 128, 256, \cdots$$

in which the last digit can be only one of the numbers

$$2, 4, 8, 6.$$

To find the number of digits in 2^p we recall that $\log 2^p = p \cdot \log 2$. From a table we find that $\log 2$ is approximately .30103, so

$$\log 2^p = p \cdot \log 2 = p \cdot .30103.$$

In case $p = 11213$, this gives

$$\log 2^{11213} = 3375.449 \cdots,$$

and from the characteristic 3375, we conclude that the number 2^r has 3376 digits. So we can say:

The largest prime presently known has 3376 digits. (Here the word "presently" is essential.) It was computed on a University of Illinois computer and the mathematics department was so proud of its achievement that its postage meter stamps it out on every letter for the whole world to admire.

2.3 Fermat Primes

There is another type of prime with a long and interesting history: the Fermat primes. These were introduced originally by Fermat (1601–1665), a French judge who, as a side line, was a distinguished mathematician. The first five Fermat primes are

$$F_0 = 2^{2^0} + 1 = 3, \quad F_1 = 2^{2^1} + 1 = 5, \quad F_2 = 2^{2^2} + 1 = 17,$$

$$F_3 = 2^{2^3} + 1 = 257, \quad F_4 = 2^{2^4} + 1 = 65537.$$

According to this sequence the general formula for the Fermat primes should be

(2.3.1) $F_n = 2^{2^n} + 1.$

Fermat firmly believed that all numbers of this kind were primes although he did not carry his calculations beyond the five numbers given above. This conjecture was thrown out of the window when the Swiss mathematician Euler went one step further and showed that the next Fermat number,

$$F_5 = 4294967297 = 641 \cdot 6700417,$$

is not a prime, as indicated. This probably would have been the end of the story had not the Fermat numbers arisen in an entirely different problem, the construction of regular polygons with straightedge and compasses.

A regular polygon is a polygon whose vertices lie at equal distances from each other on a circle (Figure 2.3.1). If a regular polygon has n vertices, we call it a *regular n-gon*. The n lines from the vertices to the center of the circle create n central angles, each of size

$$\frac{1}{n} \cdot 360°.$$

If one can construct an angle of this size one can also construct the n-gon.

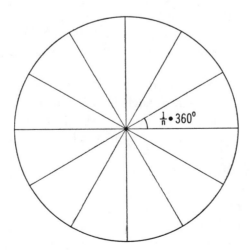

Figure 2.3.1

The ancient Greeks were much interested in finding methods for constructing regular polygons with compasses and straightedge. The simplest cases of an equilateral triangle and a square they could, of course, construct. By repeatedly taking half the central angle, they could therefore construct regular polygons with

$$4, 8, 16, 32, \cdots,$$

$$3, 6, 12, 24, \cdots$$

vertices. Furthermore, they could construct the regular pentagon; hence also the regular polygons with

$$5, 10, 20, 40, \cdots$$

vertices. One further type of regular polygons was obtainable. The central angle in a 15-gon is

$$\tfrac{1}{15} \cdot 360° = 24°,$$

and this can be derived from the angle 72° in the pentagon and the angle 120° in the 3-gon: one takes the first angle twice and subtracts the second. Hence we can construct regular polygons of 15, 30, 60, 120, $\cdots$ sides.

This was the state of affairs until 1801, when the young German mathematician C. F. Gauss (1777–1855) published an epoch-making work on number theory: *Disquisitiones Arithmeticae* (English edition, Yale University Press, 1966). Not only did Gauss surpass the Greek geometers by giving a construction by compasses and straightedge for the regular 17-gon, but he went much further. He determined, for all n, which n-gons can be so constructed and which cannot. We shall now describe Gauss's results.

We noticed above that from a regular n-gon one can obtain a $2n$-gon by dividing each central angle in half. On the other hand, from a $2n$-gon one can construct an n-gon simply by using every other vertex. This shows that to determine which regular polygons can be constructed it suffices to examine only the n-gons with odd n. For such n Gauss showed: *A regular polygon with n vertices can be constructed by compasses and straightedge if, and only if, n is a Fermat prime or a product of distinct Fermat primes.*

Let us examine the smallest values of n. One sees that 3-gons and 5-gons can be constructed, while a 7-gon cannot, since 7 is not a Fermat prime. A 9-gon cannot be constructed, since $9 = 3 \cdot 3$ is the product of two equal Fermat primes. For $n = 11$ or $n = 13$ the polygon cannot be constructed, but it can be constructed for $n = 15 = 3 \cdot 5$ and $n = 17$.

Gauss's discovery naturally created new interest in the Fermat numbers (2.3.1). In the last century many heroic calculations were made, unaided by machines, to find new Fermat primes. At present these calculations go on at an accelerated rate by means of computers. So far the results have been negative. No new Fermat primes have been found and many mathematicians are now inclined to believe that there are no more of them.

Problem Set 2.3

1. Find all odd $n < 100$ for which a regular n-gon can be constructed.

2. How would you construct a regular polygon with 51 sides assuming that you know one with 17 sides?

3. If there are no Fermat primes other than the five mentioned, how many regular n-gons (n odd) can be constructed? What is the largest odd n for which a regular n-gon can be constructed?

2.4 The Sieve of Eratosthenes

As we have stated, there are tables of primes up to quite large numbers. How should one go about actually constructing such a table? This problem was solved, in a way, by the Alexandrian mathematician Eratosthenes (about 200 B.C.). His scheme runs as follows: We write the sequence of all integers from 1 upward to whatever point we may wish to go:

1	2	3	4	5	6	7	8	9	10	11	12	13	14	15
			2		2		2	3	2		2		2	3

We start with the prime 2. We eliminate every second number after 2 (but not 2 itself), that is, the even numbers 4, 6, 8, 10, etc., by placing a mark under each. After this has been done the first unmarked number is 3. It is a prime since it is not divisible by 2. We leave 3 unmarked but mark every third number after it, that is, the numbers 6, 9, 12, 15, $\cdots$; some of them already have been marked since they are even. In the next step the first unmarked number is 5; it is a prime since it is not divisible by 2 or 3. We leave 5 unmarked, but mark every fifth number afterward, that is, the numbers 10, 15, 20, 25, $\cdots$ which have not already been marked. Now the smallest unmarked number is 7; it is a prime since it is not divisible by any of the smaller primes 2, 3, 5. By repeating this process one finally ends up with a sequence of unmarked numbers; these (with the exception of the number 1) are the primes up to the given point.

This method of sifting the numbers is known as the *sieve of Eratosthenes*. Every prime table is constructed from this sieve principle. Actually one can go much further by using the memory of computers. At the Los Alamos Scientific Laboratory all primes up to 100,000,000 have been stored in this way.

Through a little variation of the sieve method one can obtain more information. Suppose that each time a number in the sequence is marked for the first time we write under the mark the prime which eliminates it. Then 15 and 35 would become

$$\frac{15}{3}, \quad \frac{35}{5}$$

and so on as shown in the sequence above. Thus we have indicated the primes, and, for each composite number, we have given the smallest prime which divides it. Such a list of numbers is called a *factor table*. A factor table is more elaborate than a prime table; to simplify it a little one usually eliminates from it the composite numbers with small prime factors like 2, 3, 5, 7. The largest factor table we have was computed by D. N. Lehmer and extends to all numbers up to 10,000,000.

The sieve of Eratosthenes can be used to construct prime tables and factor tables as we have seen. But it can also be used for theoretical purposes, and many important results in modern number theory have been derived by means of the sieve method. Let us point out one fact already known to Euclid:

There is an infinite number of primes.

PROOF: Suppose there were only k primes

$$2, 3, 5, \cdots, p_k.$$

Then in the sieve there would be no unmarked numbers after p_k. But this is impossible. For, the product of the primes

$$P = 2 \cdot 3 \cdot 5 \cdots p_k$$

would be eliminated k times, once for each prime, so the next number $P + 1$ cannot be marked for any of them.

Problem Set 2.4

1. Tabulate the number of primes for each hundred: 1–100, 101–200, up to 901–1000.

2. Try to determine the number of primes in the range 10 001 – 10 100.

CHAPTER THREE

Divisors of Numbers

3.1 Fundamental Factorization Theorem

A composite number c can be written as a product $c = a \cdot b$, where neither of the factors is 1 and both are less than c; for example,

$$72 = 8 \cdot 9, \qquad 150 = 10 \cdot 15.$$

In the factorization of c, one or both of the factors a and b may be composite. If a is composite, it can be factored further:

$$a = a_1 \cdot a_2, \qquad c = a_1 \cdot a_2 \cdot b.$$

In the examples above,

$$72 = 2 \cdot 4 \cdot 9, \qquad 150 = 2 \cdot 5 \cdot 15.$$

One can continue this process of factorization until it stops; it must stop because the factors become smaller and smaller, but cannot be 1. When one can factor no further, every factor is a prime. Thus we have shown:

Every integer greater than 1 is a prime or a product of primes.

The stepwise factorization of a number can be accomplished in many ways. One could use a factor table and first find the smallest prime p_1 which divides c so that $c = p_1 \cdot c_1$. If c_1 is composite the table gives the smallest prime p_2 dividing c_1 so that

$$c_1 = p_2 \cdot c_2, \qquad c = p_1 \cdot p_2 \cdot c_2.$$

Then one finds the smallest prime factor of c_2, and so on.

But it is a fundamental fact that, regardless of how the factorization into primes is carried out, the result is always the same except for the order of the factors; that is, in any two prime factorizations the primes are the same and each occurs the same number of times. This result we express briefly by saying:

The prime factorization of a number is unique.

Perhaps you have heard and used this so-called "fundamental theorem of arithmetic" so often that you feel it is pretty obvious; this is not so. The theorem can be proved in various ways, but none of them is trivial. Here we shall employ a method of proof which is known by the Latin term: *reductio ad absurdum*. This means that one assumes that the theorem to be proved is false and shows that this leads to an absurd result.

PROOF: Suppose that our unique factorization theorem is not true. Then there exist numbers with more than one prime factorization. Among these there must be a smallest one which we shall call c_0. The theorem is true for small integers, say up to 10, as one sees by checking. The number c_0 has a smallest prime factor p_0, and we may write

$$c_0 = p_0 \cdot d_0.$$

Since $d_0 < c_0$, there is a unique prime factorization of d_0, and this means that the prime factorization of c_0 in which p_0 occurs is unique.

Since by assumption there are at least two prime factorizations of c_0, there must be one in which p_0 does not occur. The smallest prime in this decomposition we call p_1 and write

(3.1.1) $$c_0 = p_1 \cdot d_1.$$

Since $p_1 > p_0$, we have $d_1 < d_0$ and hence also $p_0 d_1 < c_0$. Let us now examine the number

(3.1.2) $$c_0' = c_0 - p_0 \cdot d_1 = (p_1 - p_0) d_1.$$

Since this is a smaller number than c_0, it must have a unique factorization, and the prime factors of c_0' are made up of the prime factors of $p_1 - p_0$ and d_1. Since c_0 is divisible by p_0, it follows from the expression (3.1.2) that c_0' is also divisible by p_0; hence p_0 must divide either d_1 or $p_1 - p_0$. But the prime factors in d_1 are greater than p_0 since p_1 was the smallest prime in the decomposition (3.1.1). Thus the only other possibility is that p_0 divides $p_1 - p_0$; hence it divides p_1. But this is absurd, because a prime p_1 cannot be divisible by another prime p_0.

We said above that it is by no means obvious that a number can be factored into primes in only one way. Indeed, there are many "arithmetics" where an analogous theorem fails to be true. To give a very simple example let us take a look at the even numbers

$$2, \ 4, \ 6, \ 8, \ 10, \ 12, \ \cdots.$$

Some of these can be factored into two even factors, some cannot. The latter we may call even-primes. They are the numbers divisible by 2 but not by 4:

$$2, \ 6, \ 10, \ 14, \ 18, \ \cdots.$$

One sees that every even number either is an even-prime or can be written as the product of even-primes. But such an even-prime factorization need not be unique; for instance, the number 420 has the different even-prime decompositions:

$$420 \ = \ 6 \cdot 70 \ = \ 10 \cdot 42 \ = \ 14 \cdot 30.$$

Problem Set 3.1

1. Find the prime factorization of each of the numbers

$$120, \quad 365, \quad 1970.$$

2. Do the same for the numbers in the problems of Section 2.1, page 17.

3. Write all even-prime factorizations of 360.

4. When does an even number have a unique even-prime factorization?

3.2 Divisors

Let us factor a number, say 3600. The factorization

$$3600 = 2 \cdot 2 \cdot 2 \cdot 2 \cdot 3 \cdot 3 \cdot 5 \cdot 5$$

can be written

$$3600 = 2^4 \cdot 3^2 \cdot 5^2.$$

Similarly, in general, when one factors a number n one can gather the equal prime factors into powers and write

(3.2.1) $$n = p_1^{\alpha_1} \cdot p_2^{\alpha_2} \cdots p_r^{\alpha_r},$$

where p_1, p_2, $\cdots$, p_r are the different prime factors of n, and p_1 occurs α_1 times, p_2 occurs α_2 times, and so on. Once we know this form (3.2.1) of a number, we can answer certain questions about the number right away.

For instance, we may want to know which numbers divide n. Take as an example the number 3600 we mentioned above. Suppose d is one of its divisors, so that

$$3600 = d \cdot d_1.$$

The prime factorization shows that the only primes which can possibly occur as factors of d are 2, 3, 5. Furthermore, 2 can occur as a factor at most 4 times while 3 and 5 can each occur at most 2 times. So we see that the possible divisors of 3600 are

$$d = 2^{\delta_1} \cdot 3^{\delta_2} \cdot 5^{\delta_3}$$

where we have the choices

$$\delta_1 = 0, 1, 2, 3, 4; \quad \delta_2 = 0, 1, 2; \quad \delta_3 = 0, 1, 2$$

for the exponents. Since these choices can be combined in all possible ways, the number of divisors is

$$(4 + 1)(2 + 1)(2 + 1) = 5 \cdot 3 \cdot 3 = 45.$$

The situation for any number n with prime factorization (3.2.1) is just the same. When d is a divisor of n, that is,

$$n = d \cdot d_1,$$

then the only primes which can divide d are those which divide n, namely $p_1, \cdots, p_r$; so we can write the prime factorization of d in the form

$$(3.2.2) \qquad d = p_1{}^{\delta_1} \cdot p_2{}^{\delta_2} \cdots p_r{}^{\delta_r}.$$

The prime p_1 can occur at most α_1 times, as in n, and similarly for p_2 and the other primes. This means that for δ_1 we have the $\alpha_1 + 1$ choices

$$\delta_1 = 0, 1, \cdots, \alpha_1$$

and similarly for the other primes. Since each of the $\alpha_1 + 1$ choices of value for δ_1 can be combined with the $\alpha_2 + 1$ possible values for δ_2 and so on, we see that the total number $\tau(n)$ of divisors of n is given by the formula

$$(3.2.3) \qquad \tau(n) = (\alpha_1 + 1)(\alpha_2 + 1) \cdots (\alpha_r + 1).$$

Problem Set 3.2

1. How many divisors has a prime? A prime power p^α?

2. Find the number of divisors of the following numbers:

$$60; \quad 366; \quad 1970; \quad \text{your Zip number.}$$

3. Which integer (or integers) up to and including 100 has the largest number of divisors?

3.3 Problems Concerning Divisors

The only number with just one divisor is $n = 1$. The numbers with exactly two divisors are the primes $n = p$; they are divisible by 1 and p. The smallest number with two divisors is therefore $p = 2$.

Let us examine the numbers with exactly 3 divisors. According to (3.2.3) we have

$$3 = (\alpha_1 + 1)(\alpha_2 + 1) \cdots (\alpha_r + 1).$$

Since 3 is a prime, there can be only one factor $\neq 1$ on the right, so

$r = 1$ and $\alpha_1 = 2$. Thus

$$n = p_1^2.$$

The smallest number with 3 divisors is $n = 2^2 = 4$. This argument applies to any case where the number of divisors is a prime q; one finds

$$q = \alpha_1 + 1, \quad \text{so} \quad \alpha_1 = q - 1 \quad \text{and} \quad n = p_1^{q-1},$$

and the smallest such number is

$$n = 2^{q-1}.$$

Consider next the case where there are 4 divisors. Then

$$4 = (\alpha_1 + 1)(\alpha_2 + 1)$$

is possible only when

$$\alpha_1 = 3, \quad \alpha_2 = 0, \quad \text{or} \quad \alpha_1 = \alpha_2 = 1.$$

This leads to two alternatives

$$n = p_1^3, \quad n = p_1 \cdot p_2,$$

and the smallest number with 4 divisors is $n = 6$.

When there are 6 divisors one has

$$6 = (\alpha_1 + 1)(\alpha_2 + 1),$$

and this is only possible when

$$\alpha_1 = 5, \quad \alpha_2 = 0 \quad \text{or} \quad \alpha_1 = 2, \quad \alpha_2 = 1.$$

This gives the alternatives

$$n = p_1^5, \quad n = p_1^2 \cdot p_2,$$

and the smallest value occurs in the latter case when

$$p_1 = 2, \quad p_2 = 3, \quad n = 12.$$

One may use this method to calculate the smallest integer with any given number of divisors.

There exist tables of the number of divisors for the various numbers. They begin as follows:

$n =$	1	2	3	4	5	6	7	8	9	10	11	12
$\tau(n) =$	1	2	2	3	2	4	2	4	3	4	2	6

and you may readily continue for yourself.

Let us say that an integer n is *highly composite* when all numbers less than n have fewer divisors than n has. By looking at our little table we see that

$$1, \quad 2, \quad 4, \quad 6, \quad 12$$

are the first among the highly composite numbers We know little about the properties of these numbers.

Problem Set ⁓ ⁓

1. A platoon of 12 soldiers can march in 6 different formations: 12×1, 6×2, 4×3, 3×4, 2×6, 1×12. What are the smallest companies of men which can march in 8, 10, 12 and 72 ways?

2. Find the smallest integers with 14 divisors, 18 divisors and 100 divisors.

3. Find the first two highly composite numbers following 12.

4. Characterize all integers for which the number of divisors is a product of two primes.

3.4 Perfect Numbers

The ancient Greeks were very fond of numerology or gematry, as it is sometimes called. A natural reason for this was that Greek numbers were expressed by means of the letters in the Greek alphabet so that each written word, each name, was associated with a number. Two men could compare the properties of the numbers of their names.

The divisors or *aliquot parts* of a number were particularly important in gematry. Most ideal, in fact *perfect*, were those numbers which were just made up of their aliquot parts, that is, the sum of the divisors was equal to the number. Here it must be noted that the Greeks did not consider the number itself to be a divisor.

The smallest perfect number is

$$6 = 1 + 2 + 3.$$

The next is

$$28 = 1 + 2 + 4 + 7 + 14,$$

and the next

$$496 = 1 + 2 + 4 + 8 + 16 + 31 + 62 + 124 + 248.$$

Often a mathematician who has one or more special solutions to a problem will play around with these trying to find some regularities that give a clue to the general solution. For our special perfect numbers we have

$$6 = 2 \cdot 3 = 2 \cdot (2^2 - 1),$$
$$28 = 2^2 \cdot 7 = 2^2(2^3 - 1),$$
$$496 = 2^4 \cdot 31 = 2^4(2^5 - 1).$$

This leads us to the educated guess:

A number is perfect when it has the form

(3.4.1) $$P = 2^{p-1}(2^p - 1) = 2^{p-1}q,$$

where

$$q = 2^p - 1$$

is a Mersenne prime.

This result was actually known to the Greeks and it is not hard to prove. The divisors of the number P including P itself are seen to be

$$1, \quad 2, \quad 2^2, \quad \cdots, \quad 2^{p-1},$$
$$q, \quad 2q, \quad 2^2q, \quad \cdots, \quad 2^{p-1}q.$$

The sum of these divisors is

$$1 + 2 + \cdots + 2^{p-1} + q(1 + 2 + \cdots + 2^{p-1})$$

which is equal to

$$(1 + 2 + \cdots + 2^{p-1})(q + 1) = (1 + 2 + \cdots + 2^{p-1})2^p.$$

In case you don't remember the sum of the geometric series

$$S = 1 + 2 + \cdots + 2^{p-1},$$

multiply it by 2

$$2S = 2 + 2^2 + \cdots + 2^{p-1} + 2^p$$

and subtract S to get

$$S = 2^p - 1 = q.$$

Thus the sum of all divisors of P is

$$2^p q = 2 \cdot 2^{p-1} q,$$

and the sum of all divisors excluding $P = 2^{p-1} q$ is

$$2 \cdot 2^{p-1} q - 2^{p-1} q = 2^{p-1} q = P,$$

so our number is perfect.

This result shows that each Mersenne prime gives rise to a perfect number. In Section 2.2 we mentioned that so far 23 Mersenne primes are known, so we also know 23 perfect numbers. Are there any other types of perfect numbers? All the perfect numbers of the form (3.4.1) are even and it is possible to prove that if a perfect number is even it must be of the form (3.4.1). This leaves us with the question: Are there any odd perfect numbers? Presently we know of none and it is one of the outstanding puzzles of number theory to determine whether an odd perfect number can exist. It would be quite an achievement to come up with one and you may be tempted to try out various odd numbers. We should advise against it; according to a recent announcement by Bryant Tuckerman at IBM (1968), an odd perfect number must have at least 36 digits.

Problem Set 3.4

1. By means of the list of Mersenne primes compute the fourth and fifth perfect numbers.

3.5 Amicable Numbers

Another bequest of Greek numerology is the amicable numbers. When two men had names with number values so related that the sum of the parts (divisors) of one was equal to the other and vice versa, this was taken to be a sign of an intimate relation between the two. Actually the Greeks knew of only a single pair of such numbers, namely

$$220 = 2^2 \cdot 5 \cdot 11, \quad 284 = 2^2 \cdot 71.$$

Their sums of divisors are respectively

$$1 + 2 + 4 + 5 + 10 + 20 + 11 + 22 + 55 + 110 = 284,$$

$$1 + 2 + 4 + 71 + 142 = 220.$$

Fermat saved the theory of amicable numbers from being based on a single instance by finding the pair

$$17296 = 2^4 \cdot 23 \cdot 47, \quad 18416 = 2^4 \cdot 1151.$$

The search for amicable pairs is eminently suited for computers. For each number n, one lets the machine determine all divisors $(\neq n)$ and their sum m. Then in a second step one performs the same operation on m. If one returns to the original number n by this operation, an amicable pair (n, m) has been discovered. A sweep of this kind was recently undertaken for all n below one million on the Yale computer IBM 7094. This resulted in the collection of 42 pairs of amicable numbers; some of them are new. Those below 100,000 are given in Table 2. The method as set up will also catch the perfect numbers. If anyone should wish to proceed beyond a million, he can certainly do so by using more computer time.

TABLE 2. AMICABLE NUMBERS UP TO 100 000.

$220 = 2^2 \cdot 5 \cdot 11$	$284 = 2^2 \cdot 71$
$1184 = 2^5 \cdot 37$	$1210 = 2 \cdot 5 \cdot 11^2$
$2620 = 2^2 \cdot 5 \cdot 131$	$2924 = 2^2 \cdot 17 \cdot 43$
$5020 = 2^3 \cdot 5 \cdot 251$	$5564 = 2^2 \cdot 13 \cdot 107$
$6232 = 2^3 \cdot 19 \cdot 41$	$6368 = 2^5 \cdot 199$
$10744 = 2^3 \cdot 17 \cdot 79$	$10856 = 2^3 \cdot 23 \cdot 59$
$12285 = 3^3 \cdot 5 \cdot 7 \cdot 13$	$14595 = 3 \cdot 5 \cdot 7 \cdot 139$
$17296 = 2^4 \cdot 23 \cdot 47$	$18416 = 2^4 \cdot 1151$
$63020 = 2^2 \cdot 5 \cdot 23 \cdot 137$	$76084 = 2^2 \cdot 23 \cdot 827$
$66928 = 2^4 \cdot 47 \cdot 89$	$66992 = 2^4 \cdot 53 \cdot 79$
$67095 = 3^3 \cdot 5 \cdot 7 \cdot 71$	$71145 = 3^3 \cdot 5 \cdot 17 \cdot 31$
$69615 = 3^2 \cdot 5 \cdot 7 \cdot 13 \cdot 17$	$87633 = 3^2 \cdot 7 \cdot 13 \cdot 107$
$79750 = 2 \cdot 5^3 \cdot 11 \cdot 29$	$88730 = 2 \cdot 5 \cdot 19 \cdot 467$

Actually we know very little about the properties of the amicable numbers, but on the basis of our table one can make some conjectures. For instance, it appears that the quotient of the two numbers must get closer and closer to 1 as they increase. From the table one sees that both numbers may be even or both odd, but no case has been found in which one was odd and the other even. A search for amicable numbers of this kind up to much higher limits has been made, but none have been found for

$$n \leqslant 3\,000\,000\,000.$$

CHAPTER FOUR

Greatest Common Divisor and Least Common Multiple

4.1 Greatest Common Divisor

We frankly hope you will find much of this chapter superfluous. It deals with concepts with which you should have become acquainted as soon as you learned to calculate with fractions in grade school. All that can be said to justify the inclusion of these things is that it may refresh your memory and that the presentation may be more systematic than what you were accustomed to.

Let us take some fraction a/b, the quotient of two integers a and b. Usually we try to reduce it to its lowest terms, that is, we seek to cancel out factors common to a and b. This operation does not change the value of the fraction; for instance,

$$\frac{24}{36} = \frac{8}{12} = \frac{2}{3}.$$

A *common divisor* d of two integers a and b is an integer d which is a factor of both a and b; that is,

$$a = d \cdot a_1, \qquad b = d \cdot b_1.$$

If d is a common divisor of a and b, it also divides the numbers $a + b$ and $a - b$ since

$$a + b = a_1 d + b_1 d = (a_1 + b_1)d,$$

$$a - b = a_1 d - b_1 d = (a_1 - b_1)d.$$

39

When one knows the prime factorizations of a and b it is not difficult to find all common divisors. We write the two prime factorizations as follows:

(4.1.1) $$a = p_1^{\alpha_1} \cdots p_r^{\alpha_r}, \qquad b = p_1^{\beta_1} \cdots p_r^{\beta_r}.$$

Here we agree to write the factorizations as if a and b had the same prime factors

$$p_1, \; p_2, \; \cdots, \; p_r,$$

but with the convention that we include the possibility of using exponents which are 0. For example, if p_1 divides a but not b, we put $\beta_1 = 0$ in (4.1.1). Thus, if

(4.1.2) $$a = 140, \qquad b = 110,$$

we write

(4.1.3) $$a = 2^2 \cdot 5^1 \cdot 7^1 \cdot 11^0, \qquad b = 2^1 \cdot 5^1 \cdot 7^0 \cdot 11^1.$$

In (4.1.1) a divisor d of a can have as prime factors only the p_i occurring in a, and each to an exponent δ_i not greater than the corresponding α_i in a. The analogous conditions hold for any divisor d of b. Therefore, a common divisor d of a and b can have as prime factors only the p_i that occur in both a and b; and the exponent δ_i of p_i in d cannot exceed the smaller of the two exponents α_i and β_i.

From this discussion we conclude:

Any two integers a and b have a *greatest common divisor* d_0. The prime factors p_i of d_0 are those which occur both in a and in b, and the exponent of p_i in d_0 is the smaller of the two numbers α_i and β_i.

Example. Take the numbers in (4.1.2) with the prime factorizations (4.1.3); one sees that

$$d_0 = 2^1 \cdot 5^1 = 10.$$

Since the exponent of a prime p_i in the greatest common divisor is at least as great as in any other common divisor we have the characteristic property:

Any common divisor d divides the greatest common divisor d_0.

The g.c.d. (greatest common divisor) of two numbers is so important that there is a special notation for it:

$$(4.1.4) \qquad d_0 = (a, b).$$

Problem Set 4.1

1. Find the g.c.d. of the pairs of numbers
   (a) 360 and 1970.
   (b) 30 and 365.
   (c) Your telephone number and your Zip number.

2. How would you prove that $\sqrt{2}$ is irrational? How does the theorem of the unique prime factorization enter into this and similar proofs?

4.2 Relatively Prime Numbers

The number 1 is a common divisor of any pair of numbers a and b. It may happen that this is the only common factor, so that

$$(4.2.1) \qquad d_0 = (a, b) = 1.$$

In this case we say that a and b are *relatively prime*.

Example: $(39, 22) = 1$.

If the numbers have a common divisor greater than 1, they also have a common prime divisor; so two numbers can only be relatively prime when they have no common prime factors. The condition (4.2.1) means, therefore, that a and b have no common primes, that is, all their prime factors are different.

Let us return to the starting point for this chapter, the reduction of a fraction a/b to its lowest terms. If d_0 is the g.c.d. of a and b

we can write

(4.2.2) $a = a_0 d_0, \quad b = b_0 d_0.$

We then have

(4.2.3) $\dfrac{a}{b} = \dfrac{a_0 d_0}{b_0 d_0} = \dfrac{a_0}{b_0}.$

In (4.2.2) there can be no common prime factor for a_0 and b_0, for otherwise a and b would have a common factor greater than d_0. We conclude that

(4.2.4) $(a_0, b_0) = 1.$

This means that the second fraction in (4.2.3) is in its lowest terms; no further cancellation is possible.

One property of relatively prime numbers often comes into play:

DIVISION RULE: *If a product ab is divisible by a number c which is relatively prime to b, then a is divisible by c.*

PROOF: Since c divides ab, the prime factors of c occur among those of a and b. But since $(b, c) = 1$, they cannot occur in b. Thus all prime factors of c divide a but not b, and they appear in a to powers which are not less than those in c since c divides ab.

Later on we shall make use of another fact:

If a product of two relatively prime numbers is a square,

(4.2.5) $ab = c^2, \quad (a, b) = 1,$

then both a and b are squares:

(4.2.6) $a = a_1^2, \quad b = b_1^2.$

PROOF: For a number to be a square it is necessary and sufficient that all exponents in the prime factorization be even. Since a and b are relatively prime in (4.2.5) any prime factor in c^2 occurs in a or in b, but not in both; so the prime factors in a and in b must have even exponents.

Problem Set 4.2

1. Which are the numbers relatively prime to 2?

2. Why is $(n, n + 1) = 1$ for two consecutive integers n and $n + 1$?

3. Check the pairs of amicable numbers in Table 2 (p. 36) and find out which of them are relatively prime.

4. Does the rule expressed in (4.2.5) and (4.2.6) hold for arbitrary powers instead of squares?

4.3 Euclid's Algorithm

Let us go back to our fractions a/b again. If $a > b$, the fraction is a number greater than 1, and we often separate it into an integral part and a proper fraction less than 1.

Examples. We write

$$\tfrac{32}{5} = 6 + \tfrac{2}{5} = 6\tfrac{2}{5}; \qquad \tfrac{63}{7} = 9 + \tfrac{0}{7} = 9.$$

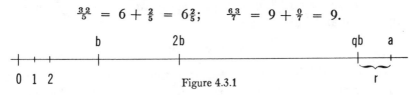

Figure 4.3.1

To do this in general, we make use of the (*incomplete*) *division* of two integers $a \geqslant b$: we write

(4.3.1) $a = qb + r$ with $0 \leqslant r \leqslant b - 1$

To see that this is always possible, we represent the integers 0, 1, 2, $\cdots$ on the number line (Figure 4.3.1). Somewhere on the line the number a is represented. Starting at 0 we mark off b, $2b$, $3b$, and so on up to qb such that qb is not greater than a, while $(q + 1)b$ is. The distance from qb to a is r. We call r the *remainder* in the division (4.3.1), and q the (*incomplete*) *quotient*. This quotient q occurs so often that there is a special symbol for it,

$$q = \left[\frac{a}{b}\right];$$

this symbol denotes *the greatest integer contained in a/b*. For the examples above we have

$$\left[\tfrac{32}{5}\right] = 6, \qquad \left[\tfrac{63}{7}\right] = 9.$$

In the preceding section we examined the g.c.d.

$$(4.3.2) \qquad\qquad d_0 = (a, b)$$

of two integers a and b. To find d_0 we assumed that we knew the prime factorizations of a and b. To determine these may be a formidable task for large numbers. There is an important, quite different method for finding the g.c.d. which does not depend on the factorizations. It is based upon the following argument:

If $a = qb + r$, with $0 \leqslant r \leqslant b - 1$, then

$$(4.3.3) \qquad\qquad (a, b) = d = (r, b).$$

PROOF: Let us write

$$d_0 = (a, b), \qquad d_1 = (r, b)$$

so that the relation (4.3.3) to be proved says $d_0 = d_1$. Any common divisor of a and b also divides

$$r = a - qb;$$

consequently d_0 divides r. Since d_0 is a divisor of r and also of b, it must divide $d_1 = (b, r)$, so that $d_1 \geqslant d_0$. On the other hand, according to (4.3.1) any common divisor of r and b divides a, so d_1 divides a. Since d_1 is also a divisor of b it must divide $d_0 = (a, b)$; consequently $d_0 \geqslant d_1$. We conclude $d_0 = d_1$.

Example. $1066 = 5 \cdot 200 + 66$; hence $(1066, 200) = (66, 200)$.

The result expressed in the rule (4.3.3) gives us a simple method for computing the g.c.d. of two numbers. Instead of looking for the g.c.d. of a and b it suffices to find the g.c.d. of r and b. This should be simpler since r is a number which is less than both a and b. To find the g.c.d. of r and b, we use the same method and divide b by r:

$$b = q_1 r + r_1,$$

where r_1 is smaller than both b and r. By the rule (4.3.3) we obtain

$$d_0 = (a, b) = (b, r) = (r, r_1).$$

Next we treat r and r_1 similarly, and so on. The result is a series of pairs of numbers, each having the same greatest common divisor:

$$(4.3.4) \quad d_0 = (a, b) = (b, r) = (r, r_1) = (r_1, r_2) = \cdots.$$

Since the remainders decrease steadily the series must end by coming to a remainder $r_{k+1} = 0$. This happens in the division

$$r_{k-1} = q_{k+1}r_k + 0,$$

so r_k divides r_{k-1}. Then

$$(r_{k-1}, r_k) = r_k,$$

and (4.3.4) shows that

$$d_0 = (a, b) = r_k.$$

In other words, d_0 is equal to the first remainder r_k that divides the remainder preceding it.

Example. Let us find the g.c.d. of the numbers 1970 and 1066. When we divide one number by the other and continue as above, we find:

$$1970 = 1 \cdot 1066 + 904$$
$$1066 = 1 \cdot 904 + 162$$
$$904 = 5 \cdot 162 + 94$$
$$162 = 1 \cdot 94 + 68$$
$$94 = 1 \cdot 68 + 26$$
$$68 = 2 \cdot 26 + 16$$
$$26 = 1 \cdot 16 + 10$$
$$16 = 1 \cdot 10 + 6$$
$$10 = 1 \cdot 6 + 4$$
$$6 = 1 \cdot 4 + 2$$
$$4 = 2 \cdot 2 + 0.$$

Consequently $(1970, 1066) = 2$.

This method for finding the g.c.d. of two numbers is called Euclid's algorithm, since the first description of it occurs in Euclid's *Elements*. It is well suited for machine computations.

Problem Set 4.3

1. Solve the problems in Section 4.1 (page 41) by means of Euclid's algorithm.

2. Find the g.c.d. of each of the first four pairs of amicable numbers. Check the results against those obtained from the prime factorizations.

3. Find the number of zeros at the end of the number
$$n! = 1 \cdot 2 \cdot 3 \cdot \cdots \cdot n.$$
Check the result in a table of factorials.

4.4 Least Common Multiple

Let us return to our fractions. To add (or subtract) two fractions
$$\frac{c}{a}, \quad \frac{d}{b},$$
we bring them to a common denominator and then add (or subtract) the numerators.

Example.
$$\tfrac{2}{15} + \tfrac{5}{9} = \tfrac{6}{45} + \tfrac{25}{45} = \tfrac{31}{45}.$$

In general, to form the sum
$$\frac{c}{a} + \frac{d}{b}$$
we must find a common multiple of a and b; that is, a number m divisible both by a and by b. One such number is evident, namely, their product $m = ab$; so we have for the sum of the fractions

$$\frac{c}{a} + \frac{d}{b} = \frac{cb}{ab} + \frac{da}{ab} = \frac{cb + da}{ab}.$$

But there are infinitely many other common multiples of a and b. Suppose again that we know the prime factorizations of the two numbers:

(4.4.1) $\qquad a = p_1^{\alpha_1} \cdots p_r^{\alpha_r}, \qquad b = p_1^{\beta_1} \cdots p_r^{\beta_r}.$

A number m which is divisible by both a and b must be divisible by each prime p_i in a and b to a power with exponent μ_i not less than the larger of the two exponents α_i and β_i. Thus among the common multiples m there is a least one

(4.4.2) $\qquad m_0 = p_1^{\mu_1} \cdots p_r^{\mu_r},$

where each exponent μ_i is equal to the larger of α_i and β_i. This shows that m_0 is the *least common multiple* (l.c.m.), and any other common multiple of a and b is divisible by m_0. For this l.c.m. there is a special notation:

(4.4.3) $\qquad m_0 = [a, b].$

Example.

$$a = 140, \qquad b = 110.$$

The prime factorizations of these numbers are $a = 2^2 \cdot 5^1 \cdot 7^1 \cdot 11^0$, and $b = 2^1 \cdot 5^1 \cdot 7^0 \cdot 11^1$; hence

$$[a, b] = 2^2 \cdot 5^1 \cdot 7^1 \cdot 11^1 = 1540.$$

There exists the following simple relation between the g.c.d. and the l.c.m.:

(4.4.4) $\qquad ab = (a, b) \cdot [a, b]$

PROOF: When we multiply the two numbers (4.4.1) we obtain

(4.4.5) $\qquad ab = p_1^{\alpha_1 + \beta_1} \cdots p_r^{\alpha_r + \beta_r}.$

As we have noticed, the exponent of a prime p_i in (a, b) is the smaller of the two numbers α_i and β_i; in $[a, b]$ it is the greater. Suppose, for instance, that $\alpha_i \leqslant \beta_i$. Then the exponent of p_i in (a, b) is α_i, and in $[a, b]$ it is β_i; hence in their product

$$(a, b) \cdot [a, b]$$

it is $\alpha_i + \beta_i$, that is, exactly the same as in the product (4.4.5). This shows that the relation (4.4.4) holds.

Example.

$$a = 140, \quad b = 110, \quad (a, b) = 10, \quad [a, b] = 1540;$$

$$ab = 140 \cdot 110 = 10 \cdot 1540 = (a, b) \cdot [a, b].$$

From the rule (4.4.4) we see that if a and b are relatively prime, then their product equals their l.c.m.; for, in this case $(a, b) = 1$, so

$$ab = [a, b].$$

Problem Set 4.4

1. Find the l.c.m. of the pairs of numbers in the Problem Set 4.1 (page 41).

2. Find the l.c.m. for each of the first four pairs of amicable numbers.

CHAPTER FIVE

The Pythagorean Problem

5.1 Preliminaries

In the introduction (Section 1.3) we mentioned one of the most ancient of number theoretical problems: To find all right triangles with integral sides, that is, to find all integral solutions of the equation

$$(5.1.1) \qquad x^2 + y^2 = z^2.$$

This problem can be solved by means of simple properties of numbers, but before we derive the solution we shall make a few preliminary observations. A set of three integers

$$(5.1.2) \qquad (x, y, z)$$

satisfying (5.1.1) is called a *Pythagorean triple*. We disregard the trivial case where one of the sides of the triangle is zero.

It is clear that if (5.1.2) is a Pythagorean triple, then any triple

$$(5.1.3) \qquad (kx, ky, kz)$$

obtained by multiplying each of the numbers by an integer k is also Pythagorean, and conversely. Thus, in searching for the solutions, it suffices to find the *primitive triangles*, where the sides have no such common factor $k > 1$ as in (5.1.3). For example,

$$(6, 8, 10), \qquad (15, 20, 25)$$

are Pythagorean triples which both result from the primitive solution $(3, 4, 5)$.

In a *primitive triple* (x, y, z) there is no common factor of all three numbers. In fact, one can make a stronger statement: *no two of the numbers in a primitive triple have a common factor*, that is,

(5.1.4) $(x, y) = 1,$ $(x, z) = 1,$ $(y, z) = 1.$

To prove this let us suppose, for instance, that x and y have a common divisor. Then they have a common prime divisor p. According to (5.1.1) p must also divide z, so (x, y, z) would not be a primitive triple. The same argument applies to the other conditions in (5.1.4).

One can say something more about the numbers in a primitive triple. We just learned that x and y cannot both be even numbers; but we can also show that x *and* y *cannot both be odd*. Suppose namely that

$$x = 2a + 1, \qquad y = 2b + 1.$$

When we square these numbers and add them, we obtain

$$\begin{aligned} x^2 + y^2 &= (2a + 1)^2 + (2b + 1)^2 \\ &= 2 + 4a + 4a^2 + 4b + 4b^2 \\ &= 2 + 4(a + a^2 + b + b^2), \end{aligned}$$

a number which is divisible by 2 but not by 4. According to (5.1.1) this means that z^2 is divisible by 2 but not by 4. But this is not possible, for if z^2 is divisible by 2, then z is divisible by 2, and so z^2 is divisible by 4.

Since one of the numbers x, y is even and the other odd, z is also odd. *We shall assume in our notation that x is even and y is odd.*

5.2 Solutions of the Pythagorean Equation

To find the primitive solutions to the Pythagorean equation (5.1.1), we write it in the form

(5.2.1) $$x^2 = z^2 - y^2 = (z + y)(z - y).$$

We recall that x is even while z and y are odd, so all three numbers

$$x, \quad z + y, \quad z - y$$

are even. Then we can divide both sides of (5.2.1) by 4 and obtain

(5.2.2) $$(\tfrac{1}{2}x)^2 = \tfrac{1}{2}(z + y) \cdot \tfrac{1}{2}(z - y).$$

Let us put

(5.2.3) $$m_1 = \tfrac{1}{2}(z + y), \qquad n_1 = \tfrac{1}{2}(z - y)$$

so (5.2.2) becomes

(5.2.4) $$(\tfrac{1}{2}x)^2 = m_1 n_1.$$

The numbers m_1 and n_1 in (5.2.3) are relatively prime. To see this, suppose that

$$d = (m_1, n_1)$$

is the g.c.d. of m_1 and n_1. Then, as we mentioned in Section 4.1, the number d must divide both integers

$$m_1 + n_1 = z, \qquad m_1 - n_1 = y.$$

But the only common divisor of z and y in a primitive triple is 1, so

(5.2.5) $$d = (m_1, n_1) = 1.$$

Since the product (5.2.4) of these two relatively prime numbers is a square, we can use the result given at the end of Section 4.2 (page 42) to conclude that the integers m_1 and n_1 are squares:

(5.2.6) $$m_1 = m^2, \quad n_1 = n^2, \quad (m, n) = 1.$$

Here we can take $m > 0$, $n > 0$ without loss of generality. We now substitute m^2 and n^2 for m_1 and n_1, respectively, in equations (5.2.3) and (5.2.4) and obtain

$$m^2 = \tfrac{1}{2}z + \tfrac{1}{2}y, \quad n^2 = \tfrac{1}{2}z - \tfrac{1}{2}y, \quad m^2 n^2 = \tfrac{1}{4}x^2,$$

so that

(5.2.7) $$x = 2mn, \quad y = m^2 - n^2, \quad z = m^2 + n^2.$$

A check shows that these three numbers always satisfy the Pythagorean relation $x^2 + y^2 = z^2$.

It remains to determine which positive integers m and n actually correspond to primitive triangles. We shall prove that the following three conditions for m and n are necessary and sufficient

(5.2.8)
    (1) $(m, n) = 1$

    (2) $m > n$

    (3) One of the numbers m and n is even, the other odd.

PROOF: We show first that, if x, y, z form a primitive triple, conditions (5.2.8) hold. We have already shown that condition (1) is a consequence of x, y, z being relatively prime. Condition (2) follows from the fact that x, y, z are positive numbers. To see that condition (3) is necessary, we note that if m and n were both odd, then according to (5.2.7) y and z would both be even, contrary to what we deduced at the end of the previous section.

Conversely, if conditions (5.2.8) are fulfilled, (5.2.7) determines a primitive triple: condition (2) assures us that x, y and z are positive.

Could any two of them have a common prime factor p? Such a prime p dividing two of them must also divide the third since they satisfy $x^2 + y^2 = z^2$. If p divides x it must divide $2mn$ according to (5.2.7); p cannot be 2 because y and z are odd according to condition (3) and (5.2.7). Suppose $p \neq 2$ is an odd prime dividing m. Then condition (1) and the expressions (5.2.7) show that p cannot divide y and z; the same argument applies if p should divide n.

Having found the necessary and sufficient conditions (5.2.8) for m and n to give a primitive triangle, we can compute all such triangles from the expressions (5.2.7). For instance, let us take

$$m = 11, \quad n = 8.$$

Our conditions are satisfied, and we find

$$x = 176, \quad y = 57, \quad z = 185.$$

In Table 3 we have given all primitive triangles x, y, z for the smallest values of m and n.

Problem Set 5.2

1. Enlarge the table to include all values $m \leqslant 10$.

2. Can two different sets of values m, n satisfying (5.2.8) give the same triangle?

3. Find all Pythagorean triangles whose hypotenuse is < 100.

TABLE 3

m n	2	3	4	5	6	7
1	4, 3, 5		8, 15, 17		12, 35, 37	
2		12, 5, 13		20, 21, 29		28, 45, 53
3			24, 7, 25			
4				40, 9, 41		56, 33, 65
5					60, 11, 61	
6						84, 13, 85

5.3 Problems Connected with Pythagorean Triangles

We have solved the problem of finding all Pythagorean triangles. Here as almost always in mathematics the solution of one problem leads to a variety of others. Often the new questions may be considerably more difficult than the original ones.

One natural question concerning primitive triangles is: When one side in a right triangle is given, how are the others determined? Take first the case where the y-side is known. According to (5.2.7),

$$(5.3.1) \qquad y = m^2 - n^2 = (m + n)(m - n),$$

where m and n are numbers satisfying conditions (5.2.8). In (5.3.1) the two factors $(m + n)$ and $(m - n)$ are relatively prime. To see this, observe that the factors

$$(5.3.2) \qquad a = m + n, \qquad b = m - n,$$

are both odd since one of the numbers m, n is odd, the other even.

If a and b had a common odd prime factor p, then p would have to divide both numbers

$$a + b = m + n + (m - n) = 2m$$

$$\ldots = m + n - (m - n) = 2n;$$

so p would have to divide both m and n. But this is impossible since $(m, n) = 1$.

Suppose now that we have such a factorization of the given odd number y into two factors

(5.3.3) $y = ab, \quad a > b, \quad (a, b) = 1.$

From (5.3.2) we obtain

(5.3.4) $m = \frac{1}{2}(a + b), \quad n = \frac{1}{2}(a - b).$

These two numbers are also relatively prime, for any common factor would divide $a = m + n$ and $b = m - n$. Furthermore, m and n cannot both be odd, for then both a and b would be divisible by 2. We conclude that m and n satisfy the conditions (5.2.8) and so define a primitive triangle where one side is $y = m^2 - n^2$.

Example. Let $y = 15$. We have two factorizations (5.3.3), namely

$$y = 15 \cdot 1 = 5 \cdot 3.$$

The first gives.

$$m = 8, \quad n = 7, \quad x = 112, \quad y = 15, \quad z = 113,$$

while the second yields

$$m = 4, \quad n = 1, \quad x = 8, \quad y = 15, \quad z = 17.$$

Next, let the x-side be given. Since either m or n is divisible by 2, we see from $x = 2mn$ that x must be divisible by 4. If one factors $\frac{1}{2}x$ into two relatively prime factors one can take the larger as m, the smaller as n.

Example. Take $x = 24$, with

$$\tfrac{1}{2}x = 12 \cdot 1 = 4 \cdot 3.$$

The first factorization gives

$$m = 12, \quad n = 1, \quad x = 24, \quad y = 143, \quad z = 145$$

and the second

$$m = 4, \quad n = 3, \quad x = 24, \quad y = 7, \quad z = 25.$$

The third and final case brings us in touch with some important problems in number theory. If z is the hypotenuse of a primitive Pythagorean triangle, then, according to (5.2.7),

$$(5.3.5) \qquad z = m^2 + n^2;$$

that is, z is the sum of the squares of numbers m and n satisfying the conditions (5.2.8).

This leads us to pose a question already solved by Fermat: When can an integer be written as the sum of two squares

$$(5.3.6) \qquad z = a^2 + b^2 ?$$

For the moment we drop all restrictions on a and b; they may have a common factor and one or both of them may be zero. Among the integers up to 10 the following are the sums of two squares:

$$0 = 0^2 + 0^2, \quad 1 = 1^2 + 0^2, \quad 2 = 1^2 + 1^2, \quad 4 = 2^2 + 0^2$$

$$5 = 2^2 + 1^2, \quad 8 = 2^2 + 2^2, \quad 9 = 3^2 + 0^2, \quad 10 = 3^2 + 1^2.$$

The remaining numbers, 3, 6 and 7, are not representable as sums of two squares.

We shall describe how one can decide whether or not a number is the sum of two squares. Unfortunately, the proofs are not simple and must be omitted here.

We consider primes first. Every prime of the form $p = 4n + 1$ is the sum of two squares; for example,

$$5 = 2^2 + 1^2, \quad 13 = 3^2 + 2^2, \quad 17 = 4^2 + 1^2, \quad 29 = 5^2 + 2^2.$$

A remarkable fact is that such a representation can be made only in a single way.

The remaining odd primes are of the form $q = 4n + 3$; hence

$$q = 3, \quad 7, \quad 11, \quad 19, \quad 23, \quad 31, \quad \cdots.$$

No such prime is the sum of two squares; in fact, no number of the form $4n + 3$ is the sum of two squares. To see this, observe that if two integers a and b are both even, then a^2 and b^2 are both divisible by 4, so $a^2 + b^2$ is divisible by 4. If both a and b are odd, say $a = 2k + 1$, $b = 2l + 1$, then

$$a^2 + b^2 = 4k^2 + 4k + 1 + 4l^2 + 4l + 1$$
$$= 4(k^2 + l^2 + k + l) + 2,$$

so $a^2 + b^2$ has remainder 2 upon division by 4. Finally, if one of the integers a, b is even and the other odd, say $a = 2k + 1$, $b = 2l$, then

$$a^2 + b^2 = 4k^2 + 4k + 1 + 4l^2$$

has remainder 1 upon division by 4. Since this exhausts all possibilities, we conclude that the sum of two squares is never of the form $4n + 3$.

To complete our examination of the primes we notice that $2 = 1^2 + 1^2$.

The test of whether or not a composite number z is the sum of two squares runs as follows: Let the prime factorization of z be

$$(5.3.7) \qquad z = p_1^{\alpha_1} p_2^{\alpha_2} \cdots .$$

Then z is the sum of two squares if and only if every p_i of the form $4n + 3$ appears with an even exponent.

Examples.

$$z = 198 = 2 \cdot 3^2 \cdot 11$$

is not the sum of two squares since 11 is of the form $4n + 3$ and occurs to the first power.

$$z = 194 = 2 \cdot 97$$

is the sum of two squares, for neither of its prime factors is of the form $4n + 3$. One finds

$$z = 13^2 + 5^2.$$

Let us return to our original problem: to determine all numbers z that can be the hypotenuses of primitive Pythagorean triangles. Such a number z must have a representation $z = m^2 + n^2$, where the numbers m and n satisfy conditions (5.2.8). Again we omit the proofs, but it can be shown that a necessary and sufficient condition for this to be the case is that all prime factors of z be of the form $p = 4n + 1$.

Examples:

1) $z = 41$.

Here one finds a single representation as the sum of two squares of the desired kind,

$$z = 5^2 + 4^2,$$

so

$$m = 5, \quad n = 4; \qquad x = 40, \quad y = 9, \quad z = 41$$

is the corresponding triangle.

2) $z = 1105 = 5 \cdot 13 \cdot 17$.

We have four representations as the sum of two squares

$$1105 = 33^2 + 4^2 = 32^2 + 9^2 = 31^2 + 12^2 = 24^2 + 23^2.$$

We leave it to the reader to find the corresponding triangles.

A variety of problems regarding the Pythagorean triangles can be solved by means of our formulas (5.2.7),

$$x = 2mn, \qquad y = m^2 - n^2, \qquad z = m^2 + n^2.$$

For instance, one may ask for the Pythagorean triangles with a given area A. If the triangle is primitive, its area is

$$(5.3.8) \qquad A = \tfrac{1}{2}xy = mn(m - n)(m + n).$$

Three of the four factors here are odd. It is not difficult to see that they are relatively prime in pairs. So to find all possible values of m and n, one can select two relatively prime odd factors $k > l$ of A and put

$$m + n = k, \qquad m - n = l$$

giving

$$m = \tfrac{1}{2}(k + l), \qquad n = \tfrac{1}{2}(k - l).$$

Then we check these values to see whether they actually satisfy (5.3.8).

It simplifies the discussion a bit to notice that only in the special case

$$m = 2, \qquad n = 1, \qquad A = 6$$

can two of the factors in (5.3.8) be equal to 1. The only way in which there can be two unit factors in (5.3.8) is when

$$n = m - n = 1$$

giving the values above.

   Example. Find all Pythagorean triangles with area $A = 360$. The prime factorization of A is

$$A = 2^3 \cdot 3^2 \cdot 5.$$

The only way of writing A as a product of four relatively prime factors is

$$A = 8 \cdot 1 \cdot 5 \cdot 9$$

so we must have $m + n = 9$. This leads to no acceptable triangle: If $m = 8$ then $n = 1$ and $m - n = 7$ does not divide A. The other alternative is $n = 8$, $m = 1$ which is excluded by the requirement $m > n$.

   This result does not exclude the possibility of non-primitive triangles with $A = 360$. The following reasoning can be used in general to determine the non-primitive triangles with a given area. If

$$dx, \quad dy, \quad dz$$

are the sides of a triangle in which the sides have the common divisor d, then its area is

$$A = \tfrac{1}{2} \cdot dx \cdot dy = d^2 mn(m - n)(m + n).$$

So d^2 is a factor of A and, if d is the g.c.d. of the sides,

$$A_0 = \frac{A}{d^2} = mn(m - n)(m + n)$$

must be the area of a primitive triangle.

   Let us carry through this argument for the case we just considered where $A = 360$. This number has three square factors:

$$d_1 = 4, \quad d_2 = 9, \quad d_3 = 36.$$

Correspondingly one finds

$$\frac{A}{d_1} = 90 = 2 \cdot 3^2 \cdot 5, \quad \frac{A}{d_2} = 40 = 2^3 \cdot 5, \quad \frac{A}{d_3} = 10 = 2 \cdot 5.$$

There is no way of writing either 40 or 10 as the product of four relatively prime factors, and there is only one way of so representing 90, namely

$$90 = 1 \cdot 2 \cdot 3^2 \cdot 5.$$

(At most one of the four factors can be 1, except in the case $m = 2$, $n = 1$, $A = 6$.) Since 9 is the largest factor we must take $m + n = 9$. But the possible choices $m = 1, 2, 5$ yield $n = 8, 7, 4$ respectively, and the condition $m > n$ eliminates all but $m = 5$, $n = 4$ for which $mn(m + n)(m - n) \neq 90$, so we conclude that there is no Pythagorean triangle, primitive or otherwise, with area $A = 360$.

There are many other questions we could ask, but let us mention only one more. The perimeter of a triangle is

(5.3.9) $$c = x + y + z;$$

for a primitive Pythagorean triangle, it is

$$c = 2mn + (m^2 - n^2) + (m^2 + n^2) = 2m(m + n).$$

We leave it to the reader to describe some method for finding all Pythagorean triangles with a given perimeter; do not omit to apply it to numerical examples.

We have solved the problem of constructing all Pythagorean triangles. This leads us to investigate more general related problems. A natural extension is to the *Heronian triangles*, named for the Greek–Alexandrian mathematician Heron. In these triangles we require as before that the sides x, y, z be integers, but we drop the condition that one angle be 90° and replace it by the condition that the area be integral. Clearly the Pythagorean triangles fall into this category.

To check whether a given triangle is Heronian, it is simplest to use Heron's formula for the area of a triangle,

$$A = \sqrt{\tfrac{1}{2}c(\tfrac{1}{2}c - x)(\tfrac{1}{2}c - y)(\tfrac{1}{2}c - z)},$$

where c is the perimeter as we defined it in (5.3.9). Although we know a considerable number of Heronian triangles, we have no general formula giving them all. Here are the first few (non-right-angled) examples:

$x =$	7	$y =$	15	$z =$	20
	9		10		17
	13		14		15
	39		41		50

We cannot leave the Pythagorean triangles without mention of one of the most famous problems of mathematics, Fermat's conjecture: For $n > 2$ there exist no positive integers x, y, z such that

$$x^n + y^n = z^n.$$

The idea came to Fermat while he was perusing a translation from the Greek of Diophantos' *Arithmetica*. This work deals mainly with problems in which the formulas for the Pythagorean triangles are applied, and Fermat made his comments in the margin.

Fermat was excited over his "discovery" and believed that he had a wonderful proof, but unfortunately the margin was too narrow to describe it. Ever since then mathematicians have wondered. Most ingenious methods have been devised to find a proof; the search has resulted in fundamental new theories in mathematics. By means of theory and computers combined, Fermat's theorem has been proved for many exponents n; presently we know that the result is true for all n such that $3 \leqslant n \leqslant 4002$.

In view of the failure of the most prominent mathematicians during the centuries to find a general proof, the prevalent opinion seems to be that Fermat, in spite of his undisputed skill, must have been the victim of a moment of self-deception. However wide his margin had been, it is unlikely that his proof would have been valid.

You are of course entitled to make your own try; but be warned that for no theorem in mathematics have there been so many wrong proofs, a few by good mathematicians and innumerable ones from cranks. Proofs of "Fermat's last theorem" continue to make their appearance in the mail of number theorists of note, most of them with letters of demand for immediate recognition and cash payment of a monetary prize now absorbed by inflation which at one time was established by a German mathematician as a reward for a correct proof.

Problem Set 5.3

1. Find all Pythagorean triangles with one side equal to 50; to 22.

2. Use the criterion for a representation of a number as the sum of two squares to determine which of the numbers

$$100, \quad 101, \quad \cdots, \quad 110$$

has such a representation. When possible find all representations. Which of these numbers can be the hypotenuse of a primitive Pythagorean triangle?

3. Are there Pythagorean triangles with the areas

$$A = 78, \quad A = 120, \quad A = 1000 ?$$

4. Find all Pythagorean triangles with the perimeters

$$c = 88; \quad c = 110.$$

CHAPTER SIX

Numeration Systems

6.1 Numbers for the Millions

All is number, taught the ancient Pythagoreans. Yet their store of numbers was exceedingly sparse in comparison with the grotesque dance of figures which surrounds us in our present everyday existence. We count and are being counted in huge numbers; we live by social security numbers, zip numbers, account numbers, telephone numbers, room numbers and house numbers. Every day sees an influx of bills and checks and charges and balances. The official budgets unhesitatingly run into billions, and reams of statistics are an accepted form of argument. These figures are whirled around in computers which analyze the principles of big business, follow the trajectories of satellites, and explore the interior of atomic nuclei at the rate of so and so many operations per nanosecond (one billionth of a second).

All of this has developed along a continuous path from the first attempts by man to systematize his numbers as soon as they became too large to be counted on the fingers. Various methods have been in use to group numbers; most of them have fallen by the wayside when they have proved inferior in competition with other systems. Our present decadic or decimal system, based on groupings by tens, is by now, fortunately, quite universally accepted; in several respects it appears to be a fortuitous convenient middle way for our dealings with numbers.

It is not necessary to describe the system in great detail to you. After the drills of the first couple of school years, we know almost instinctively for the rest of our lives what a series of digits means, for instance

$$75 = 7 \cdot 10 + 5,$$

$$1066 = 1 \cdot 10^3 + 0 \cdot 10^2 + 6 \cdot 10 + 6,$$

$$1970 = 1 \cdot 10^3 + 9 \cdot 10^2 + 7 \cdot 10 + 0.$$

In general, in the system with base number 10 the sequence

(6.1.1) $$a_n a_{n-1} \cdots a_2 a_1 a_0$$

denotes the number

(6.1.2) $$N = a_n \cdot 10^n + a_{n-1} \cdot 10^{n-1} + \cdots + a_2 \cdot 10^2 + a_1 \cdot 10 + a_0,$$

where the coefficients or digits a_i have one of the values

(6.1.3) $$a_i = 0, 1, \cdots, 9.$$

The number $b = 10$ is called the *base* for the system.

This Hindu–Arabic number system came to Europe from the East around 1200 A.D. and has been unchallenged ever since. It is known as a positional system since the place of any digit determines its value; it is made possible by the innocuous, but ingenious, use of the symbol 0 to denote a vacant place. Moreover it has proved to be very efficient in performing our arithmetical number operations: addition, subtraction, multiplication and division.

6.2 Other Systems

There exists a large body of information about the many systems various peoples of the world have used to organize their numbers. But why and how these systems have originated is mostly lost in the nebulous past of the human race.

No one doubts that the widespread use of groupings by tens is due to the fact that humans count on their fingers. Strangely enough, there are few traces of counting on a single hand; five-systems only rarely occur. On the other foot, instances of twenty-systems are very common, and it takes no great ingenuity to visualize that this must be due to the toes also being drawn into the counting process. The Mayan count is perhaps the best known of these twenty or vigesimal systems, but they were also quite widespread in Europe until a few centuries ago. The 20-count in French from 80 to 100 is familiar, as you may recall from instances such as

$$80 = quatre\text{-}vingts,$$

$$90 = quatre\text{-}vingt\text{-}dix,$$

$$91 = quatre\text{-}vingt\text{-}onze,$$

and so on.

Less familiar, probably, is the fact that the count by scores flourishes in Danish to this very day. This ancient system, previously much more widespread among Germanic peoples, is so quaint that we cannot refrain from giving a few details. When one counts by 20's it is quite natural to use such terms as

$$tredsindstyve = \text{three times twenty,}$$

$$firsindstyve = \text{four times twenty,}$$

$$femsindstyve = \text{five times twenty.}$$

But the system becomes more complicated by the convention that whenever one has counted up to a certain number of complete 20's, and then 10 more, one states that one is half on the next twenty; for example,

$$90 = halvfemsindstyve = \text{half on the fifth twenty.}$$

To top it off, Danish uses the principle of naming the units before the tens so this results in a number construction like

$$93 = treoghalvfemsindstyve = \text{three and half on the fifth twenty.}$$

Clearly such systems are doomed in a number-infested civilization like ours. A particularly obnoxious feature of some ways of counting is to give the units before the tens. It was prevalent in English until the eighteenth century; one would say three and twenty for twenty-three. A few years ago the Norwegian Parliament abolished the system by law in school instruction and all official announcements. It still flourishes in German and is the cause of numerous number errors, for instance, in telephone dialing.

The venerable Babylonian sexagesimal (base 60) system has been in use by astronomers from antiquity till the present, although its favor is now declining. We still retain it when counting angles and time in minutes and seconds. Why the Babylonians introduced such a large basis we don't know; one guess is that it originated as a combination of two systems with different bases, say, for example, 10 and 12 with the l.c.m. 60.

It is time to say a few words about the mathematical questions involved in the use of various basis systems. With a base b, we write an integer N as

$$(6.2.1) \qquad N = c_n b^n + c_{n-1} b^{n-1} + \cdots + c_2 b^2 + c_1 b + c_0,$$

just as in (6.1.2), except that here the coefficients c_i can take the values

$$(6.2.2) \qquad\qquad c_i = 0, 1, \cdots, b - 1$$

instead of the values in (6.1.3). For short, we can write the number N in (6.2.1) in the abbreviated form

$$(6.2.3) \qquad\qquad (c_n, c_{n-1}, \cdots, c_2, c_1, c_0)_b$$

corresponding to (6.1.1); only in (6.2.3) it is necessary to attach the basis number b to avoid confusion.

Examples:

In the sexagesimal system one has

$$(3, 11, 43)_{60} = 3 \cdot 60^2 + 11 \cdot 60 + 43 = 11503.$$

In a system with base $b = 4$

$$(3, 2, 0, 1)_4 = 3 \cdot 4^3 + 2 \cdot 4^2 + 0 \cdot 4 + 1 = 225.$$

In general, when a number is given in a system with base b as in (6.2.1), one finds the number in the ordinary decimal system by computing the values of the powers of b, multiplying each by the corresponding digit, and adding, as we did in the examples above.

Next let us consider the reverse question. A number N is given and we wish to represent it in the base b. We can do this by repeated divisions by b. Take a look at the formula (6.2.1). We may write it as

$$N = (c_n b^{n-1} + \cdots + c_2 b + c_1)b + c_0.$$

Since c_0 is less than b, c_0 is the remainder obtained when N is divided by b. We can write this division

$$N = q_1 b + c_0, \qquad q_1 = c_n b^{n-1} + \cdots + c_2 b + c_1$$

to show that one obtains c_1 by dividing q_1 by b in the same way, and so on. Thus we find the coefficients c_i by a series of divisions by b:

$$N = q_1 b + c_0$$

$$q_1 = q_2 b + c_1$$

$$\cdots\cdots\cdots\cdots\cdots$$

$$q_{n-1} = q_n b + c_{n-1}$$

$$q_n = 0 \cdot b + c_n,$$

where we continue the division until $q_n < b$, $q_{n+1} = 0$, as indicated. A couple of examples will make the process clear.

Example 1. Express the number 101 in the base 3. We perform the divisions by 3 as above and find

$$101 = 33 \cdot 3 + 2$$

$$33 = 11 \cdot 3 + 0$$

$$11 = 3 \cdot 3 + 2$$

$$3 = 1 \cdot 3 + 0$$

$$1 = 0 \cdot 3 + 1.$$

This gives

$$101 = (\,1,\ 0,\ 2,\ 0,\ 2\,)_3.$$

Example 2. Express 1970 in the base 12. Here the divisions by 12 are

$$1970 = 164 \cdot 12 + 2$$

$$164 = 13 \cdot 12 + 8$$

$$13 = 1 \cdot 12 + 1$$

$$1 = 0 \cdot 12 + 1.$$

Hence

$$1970 = (\,1,\ 1,\ 8,\ 2\,)_{12}.$$

Problem Set 6.2

1. Express the numbers

$$(1, \ 2, \ 3, \ 4)_5 ; \quad (1, \ 1, \ 1, \ 1, \ 1, \ 1)_3$$

in the decimal system.

2. Represent the numbers   362;   1969;   10 000   in the bases $b = 2; \ 6; \ 17.$

6.3 Comparison of Numeration Systems

The Duodecimal Society of America has for its avowed purpose the change of our decimal number system to a supposedly more effective and convenient system with base 12. The proponents point out that it would be advantageous to have a system with a base divisible by the integers 2, 3, 4, and 6, making the division by these often recurring divisors simpler. An extension of this argument would lead us to the sexagesimal system where the base 60 is divisible by the small integers

$$2, \ 3, \ 4, \ 5, \ 6, \ 10, \ 12, \ 15, \ 20, \ 30.$$

Many things are still counted in dozens and gross (i.e. dozens of dozens), and a 12-system would certainly be feasible. One would have to introduce twelve new symbols for the digits and operate on them pretty much as we do in the decimal system. Some enthusiasts say that it is only necessary to introduce new symbols for 10 and 11; this fails to take into account the transition period where no one would know, for instance, whether 325 should mean

$$3 \cdot 10^2 + 2 \cdot 10 + 5 \ = \ 325$$

or

$$3 \cdot 12^2 + 2 \cdot 12 + 5 \ = \ 461.$$

To get a rough idea of how the number of digits in a number changes from one system to another let us take the number

$$(6.3.1) \qquad 10^n - 1 \ = \ \overset{n}{\overbrace{99 \cdots 9}} \ = \ N$$

in the decimal system. It is the largest number with n digits. To find the number of digits m which it has in the base b, we must determine m as the integer with

$$(6.3.2) \qquad b^m > 10^n - 1 \geqslant b^{m-1}.$$

This condition can also be written

$$b^m \geqslant 10^n > b^{m-1}.$$

We take the logarithms of these three numbers; when we recall that $\log 10 = 1$ it follows that

$$m \log b \geqslant n > (m - 1) \log b.$$

This in turn may be written

$$(6.3.3) \qquad m \geqslant \frac{n}{\log b} > m - 1,$$

so m is the first integer equal to or following

$$(6.3.4) \qquad \frac{n}{\log b}.$$

We conclude that, roughly, the new number of digits m is obtained by dividing n by $\log b$.

Examples. Again, let n be the number of digits of a number in the decimal system. For $b = 2$ we have $\log 2 \approx .30103$, so the number of digits in the binary system is near $3.32n$. When $b = 60$ we have $\log 60 \approx 1.778$, so the number of digits is near $.56n$, that is, a little more than half the number of digits in the decimal system.

Clearly it should be an advantage to operate with short numbers. But on the other hand a large basis number has serious disadvantages. First, one should have names and notations for the b separate digits. This is usually not done for large b. For example, in the Babylonian sexagesimal system one counted the units up to 60 in groups of ten, as illustrated in Figure 6.3.1.

$$37 = \qquad\qquad = 3 \cdot 10 + 7$$

Figure 6.3.1

This actually means that the system has been split into subsystems counted in the decimal system. A similar situation exists in the Mayan 20-system. Here the digits up to 20 were counted in 5's, as ndicated in Figure 6.3.2.

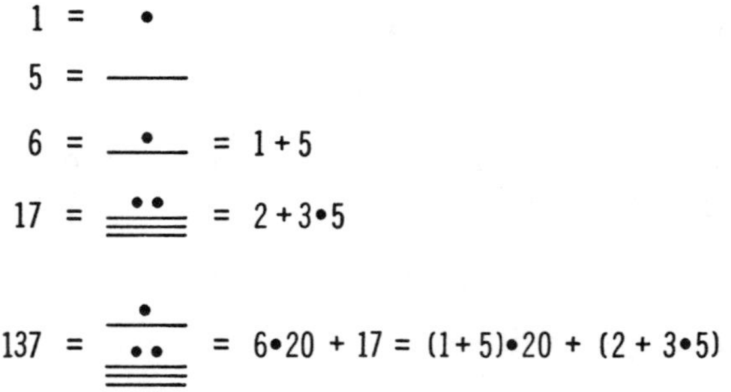

$$1 = \bullet$$

$$5 = \underline{\quad\quad}$$

$$6 = \underline{\ \bullet\ } = 1 + 5$$

$$17 = \underline{\overset{\bullet\ \bullet}{\equiv}} = 2 + 3 \bullet 5$$

$$137 = \overset{\bullet}{\underset{\underset{\equiv}{\bullet\ \bullet}}{\quad}} = 6 \bullet 20 + 17 = (1 + 5) \bullet 20 + (2 + 3 \bullet 5)$$

Figure 6.3.2

But, secondly, much greater difficulties appear when one begins to perform calculations along the usual lines. To multiply we rely upon the fact that we know by heart our multiplication table, that is, all the products of the ten digits. This Pythagorean table, as it is called in many lands, was drilled into us in the first school years so that it is nearly automatic. This knowledge is not as trivial as we may be inclined to think. From medieval arithmetical manuscripts one sees clearly that multiplication bordered upon higher mathematics and long division was a rare skill indeed. But we can take much later examples.

Samuel Pepys, of diary fame, was close to thirty years old and a clerk of the privy seal when in the summer of 1662 he decided that to check accounts independently he would have to know some mathematics, at least the fundamentals of arithmetic. At the time he had already received his bachelor's degree and master's degree from Cambridge, but it was no unusual phenomenon for a well educated British gentleman to be entirely unfamiliar with everyday reckonings; these tasks could be left to underling bookkeepers.

On July 4, 1662, Pepys writes in his diary: "By and by comes Mr. Cooper, mate of the 'Royall Charles' of whom I intend to learn mathematiques, and do begin with him today, he being a very able man, and no great matter, I suppose, will content him. After an hour's being with him at arithmetique (my first attempt being to learn the multiplication table), then we parted till tomorrow."

Pepys struggled along daily with his seaman tutor, early in the morning and late at night, to learn the confounded multiplication table. For instance, on July 9th: "Up at four o'clock and at my multiplication table hard which is all the trouble I meet withal in my arithmetique." The following days run in the same way until on July 11th he can report success: "Up at four o'clock and hard at my multiplication table which I am now almost master of." Pepys made good use of his newly won knowledge in the increasingly important positions to which he was appointed, but it may seem a bit too accelerated a progress that he was made a member of Britain's illustrious science academy, the Royal Society, two and a half years after he had learned the multiplication table.

We have inserted this little tale, which is by no means unique, to emphasize that learning the multiplication table was in earlier days no trivial step in mathematical knowledge. Thus there is much advantage, both mental and mechanical, in the use of small basis numbers in our arithmetic. For example, when the basis is $b = 3$, there is only a single non-trivial multiplication, namely

$$2 \cdot 2 = 4 = (1, 1)_3,$$

in the multiplication table

	0	1	2
0	0	0	0
1	0	1	2
2	0	2	$(1,1)_3$

For $b = 2$, we have the perfectly trivial table

	0	1
0	0	0
1	0	1 .

Problem Set 6.3

1. Prove that the number of non-trivial multiplications of the digits (that is, omitting multiplications by 0 and 1) is

$$\tfrac{1}{2}(b - 1)(b - 2)$$

in a number system with base b.

2. What is the sum of all terms in a multiplication table? Check for $b = 10$.

6.4 Some Problems Concerning Numeration Systems

Let us discuss a few problems concerning numeration systems which have some bearing upon the choice of bases for machine computation. Suppose we deal with an ordinary desk calculator that works with intermeshing number wheels, each bearing the 10 digits 0, 1, $\cdots$, 9. If there are n wheels we can represent all numbers up to

$$(6.4.1) \qquad N = \overbrace{99\cdots9}^{n}$$

as in (6.3.1).

Suppose now we use the basis number b, instead of 10, but continue to consider numbers up to N. Then we must have m wheels, where m is the integer satisfying (6.3.2) and (6.3.3), page 69. As in (6.3.4), m is the integer equal to or following the number

$$\frac{n}{\log b}.$$

Since each wheel carries b digits, the total number of digits inscribed on the wheels is approximately

$$(6.4.2) \qquad D = n \cdot \frac{b}{\log b}.$$

We may now ask: For which choice of b does one obtain the smallest number of inscribed digits? To find the smallest value of the number D in (6.4.2), one need only examine the function

$$(6.4.3) \qquad f(b) = \frac{b}{\log b}$$

for the various bases $b = 2, 3, 4, \cdots$. From a table of logarithms we get the values

b	2	3	4	5	6
$f(b)$	6.64	6.29	6.64	7.15	7.71

Succeeding values of $f(b)$ are still greater; for instance, for $b = 10$ we have $f(10) = 10$, as we have already noticed. We conclude from these calculations:

The minimal total number of digits in the calculator occurs for $b = 3$.

We also see that for $b = 2$ and $b = 4$ the total number is not much greater; so in this respect the small basis numbers have an advantage.

Let us consider a slight variation of this problem. An ordinary abacus of the type sometimes used to teach children to count has a certain number of metal wires with nine movable beads on each, to mark the digits of the numbers. One could just as well make parallel lines on paper and mark the digits by a corresponding number of matches; or one could, as in ancient sand reckoning, draw the lines on the ground and mark the digits by pebbles.

Let us stick to the abacus. If it has n wires, each with 9 beads, one can again represent all integers with n digits up to the number N in (6.4.1). We now raise the following question: Can one, by taking another basis number b, make the abacus more compact, that is, get along with a smaller number of beads?

For a basis number b the number of beads on each wire will be $b - 1$. As before, for the abacus to have the same capacity N, the number of digits or wires must be determined by (6.3.4). This yields as an approximation for the total number of beads

$$(6.4.4) \qquad\qquad E = \frac{n}{\log b} \cdot (b - 1).$$

To find out when this number has the smallest possible value, we must investigate the function

$$(6.4.5) \qquad\qquad g(b) = \frac{b - 1}{\log b}$$

for the various values $b = 2, 3, \cdots$. The values of $g(b)$ for small values of b are given in the following table:

b	2	3	4	5	6
$g(b)$	3.32	4.19	4.98	5.72	6.43

For larger values of b the function values continue to increase, so we conclude:

The number of beads required in an abacus is smallest when $b = 2$.

We can interpret this result from another point of view. Suppose we mark the digits of our number by means of matches or pebbles placed on lines. In the decimal system there will be from 0 to 9 marks on each line. This gives an average of $4\frac{1}{2}$ matches for each line when we take numbers at random; hence the numbers with n digits will on the average require $4\frac{1}{2} \cdot n$ matches when they are put in at random.

Let us examine the time it takes to put the matches in position. To have some definite figure in mind, assume that it takes one second to place a match. The total time required to place a complete number will then on the average be about $4\frac{1}{2} \cdot n$ seconds.

Suppose that we change our basis to b and assume the same capacity for representing numbers. Then there are from 0 to $b - 1$ matches on each line, hence on the average

$$\tfrac{1}{2}(b - 1)$$

of them. As we have mentioned several times, there will be approximately

$$\frac{n}{\log b}$$

lines. We conclude that the average time for marking a number with n digits is about

$$\frac{n}{\log b} \cdot \tfrac{1}{2}(b - 1) = \tfrac{1}{2}E$$

seconds where E is the expression in (6.4.4). Since this was minimal for $b = 2$ we conclude also here:

One needs the least amount of time, on the average, to place a number in position when $b = 2$.

1. Sketch the graph of the functions $y = f(b)$ in (6.4.3) and $y = g(b)$ in (6.4.5) for $b > 1$. If you are familiar with differential calculus, use it to determine the shape of the curves.

6.5 Computers and Their Numeration Systems

Until the advent of electronic computers, the decimal system reigned supreme in all fields of numerical calculations. The interest expressed in regard to other systems was mainly historical and cultural. Only a few isolated problems in mathematics could best be stated by means of binary or ternary numeration systems; one of the favorite examples in books on number theory was the game of Nim.

When the computers gradually evolved in many forms, it became essential to construct the "hardware" so that the machine could be as efficient and compact as possible. This brought on a scrutiny of numeration systems to determine the most suitable one. For many reasons, some of which we have discussed in the preceding section, the binary system was the favored candidate. In fact, its principal drawback is that we have been brought up in a different heritage, and for most of us it requires no little initial effort to feel at home in the binary system. Consequently, since the numbers which are to be coded into the computers usually come in decimal form, an initial mechanism is required to change them into binary numbers; and, in the end, the answers must be expressed in decimal form as a concession to the less mathematically trained members of the public.

The binary system used in computers is, of course, the same one we have discussed in the preceding section, but the terminology employed is prone to be more technical. The binary digits 0, 1 are called *bits*, which is short for BInary digiTS. Also, since there are only two possibilities, 0 and 1, at each position, one often talks about a two-state device.

When one follows the general rule explained in Section 6.2, the expansion of a given number in the binary system is quite simple. As an example, let us take $N = 1971$. Repeated division by

$b = 2$ yields

$$1971 = 985 \cdot 2 + 1,$$
$$985 = 492 \cdot 2 + 1,$$
$$492 = 246 \cdot 2 + 0,$$
$$246 = 123 \cdot 2 + 0,$$
$$123 = 61 \cdot 2 + 1,$$
$$61 = 30 \cdot 2 + 1,$$
$$30 = 15 \cdot 2 + 0,$$
$$15 = 7 \cdot 2 + 1,$$
$$7 = 3 \cdot 2 + 1,$$
$$3 = 1 \cdot 2 + 1,$$
$$1 = 0 \cdot 2 + 1.$$

Consequently

$$1971_{10} = (1, 1, 1, 1, 0, 1, 1, 0, 0, 1, 1)_2.$$

We noted previously that in the binary system, numbers have longer expressions; hence it becomes more difficult to size them up at a glance. For this reason the computer language often uses the *octal system* (base 8). This is only a slight variation of the binary system, obtained by dividing the bits in a number into sections of three. One can conceive of it as a system with base

$$b = 8 = 2^3 ;$$

the coefficients are the eight numbers

$$
\begin{array}{ll}
0 = 000 & 4 = 100 \\
1 = 001 & 5 = 101 \\
2 = 010 & 6 = 110 \\
3 = 011 & 7 = 111
\end{array}
$$

As an illustration, let us take the number 1971 in the example above; in the octal system it becomes

$$1971 = 011; 110; 110; 011 = (3; 6; 6; 3)_8.$$

This, as one sees, is no more than a trivially different way of writing the number. Actually we are quite familiar with it from our ordinary decimal numbers; in writing and pronouncing a large number, we usually divide the digits into groups of three, for instance,

$$N \ = \ 89 \quad 747 \quad 321 \quad 924.$$

In reality, we can say that this is a representation of our number to the base

$$b \ = \ 1000 \ = \ 10^3.$$

Other number representations are sometimes useful in computers. Suppose one wishes to record a decimal number, say $N = 2947$, in a machine based upon binary numbers. Then, instead of changing N entirely into a binary number, one could code only the digits

$$2 \ = \ 0010$$
$$9 \ = \ 1001$$
$$4 \ = \ 0100$$
$$7 \ = \ 0111$$

and so record N as

$$N \ = \ 0010; \quad 1001; \quad 0100; \quad 0111.$$

Such numbers are known as *coded decimals*. This method is sometimes called the 8421 *system*, since the decimal digits are represented as sums of the binary units

$$0 = 0000, \ 1 = 0001, \ 2 = 0010, \ 2^2 = 4 = 0100, \ 2^3 = 8 \ = \ 1000$$

These coded decimals are inconvenient for any kind of numerical calculations, but this is not always the purpose of the machine. In the same way, any letter of the alphabet and any other symbol can be assigned to some binary number. This means that any word or sentence can be kept in memory as a binary number. So if we were properly trained, and had an equally proficient audience, we could converse entirely in bits.

Problem Set 6.5

1. Find the binary expansions of the Fermat numbers (Section 2.3, page 20)

$$F_t = 2^{2^t} + 1.$$

2. Find binary expansions for the even perfect numbers (Section 3.4, page 34)

$$P = 2^{p-1}(2^p - 1).$$

6.6 Games with Digits

There are many types of games with numbers; some of them go back to medieval times. Most of them have little theoretical importance for number theory; rather, they are, like the magic squares, in the class of crossword puzzles with numbers. We shall illustrate a few of them by examples.

Consider the urgent plea of a college boy in his wire home:

$$\begin{array}{c} S\;E\;N\;D \\ M\;O\;R\;E \\ \hline M\;O\;N\;E\;Y\,. \end{array}$$

Let us conceive of this scheme as an addition of two 4-digit numbers, S E N D and M O R E, giving the sum M O N E Y. Each letter signifies a distinct digit. The problem is to determine what these digits can be. Since there are only 10 digits, there can be at most 10 different letters in each such problem; in the example above there are 8. In the ideal form, the problem should have a single solution.

In our example above, we must have

$$M = 1$$

since M is the first digit of either $S + M$ or $S + M + 1$, where S and M are numbers not larger than 9. There are then two alternatives for the number S: Since either $S + 1$ or $S + 1 + 1$ is a two digit number, the only candidates are

$$S = 9 \quad \text{or} \quad S = 8.$$

We establish first that S cannot be 8; for, if S were 8, there would have to be a carry-over from the hundreds' column to yield

$$S + M + 1 = 8 + 1 + 1 = 10$$

in the addition in the thousands' column. Consequently O would have to be zero and our message would read

$$\begin{array}{r} 8\,E\,N\,D \\ 1\,0\ R\,E \\ \hline 1\,0\,N\,E\,Y \end{array}$$

But by examining the hundreds' column, we find that there must be a carry-over from the tens' column (otherwise $E + 0 = E$, not N), and since $E \leqslant 9$, this would yield

$$E + 0 + 1 = 10.$$

This would force us to set $N = 0$; but we already have $O = 0$, so this cannot be. We conclude that

$$S = 9,$$

and the message now reads

$$\begin{array}{r} 9\,E\,N\,D \\ 1\,0\ R\,E \\ \hline 1\,0\,N\,E\,Y\,. \end{array}$$

Since $E \neq N$, the addition in the hundreds' column leads to

$$E + 1 = N,$$

and we have the situation

$$\begin{array}{ccccc} 9 & E & E+1 & D \\ 1 & 0 & R & E \\ \hline 1 & 0 & E+1 & E & Y\,. \end{array}$$

The addition in the tens' column is either

$$E + 1 + R = 10 + E \quad\text{or}\quad E + 1 + R + 1 = 10 + E.$$

The first alternative is impossible because it gives $R = 9$, contradicting $S = 9$. In the second case

$$R = 8$$

and the message reads

$$\begin{array}{ccccc}
9 & E & E+1 & D \\
1 & 0 & 8 & E \\
\hline
1 & 0 & E+1 & E & Y
\end{array}.$$

Finally, the sum in the units' column is

$$D + E = 10 + Y.$$

For the three letters D, E, Y, only the values 2, 3, 4, 5, 6, 7 are available. The sum of two different ones among these is at most 13, so we have only the alternatives $Y = 2$ or $Y = 3$. The latter alternative is not possible: it would give $D + E = 13$, and we cannot have $E = 7$, for then $N = E + 1 = 8 = R$; nor can we have $D = 7$, for then $E = 6$ and

$$N = E + 1 = 7 = D.$$

Thus we have $Y = 2$ and $D + E = 12$. Of the available digits 2, 3, 4, 5, 6, 7, the only two whose sum is 12 are 5 and 7. Since $E \neq 7$, this implies $D = 7$, $E = 5$, and so the unique solution to our problem is

$$\begin{array}{cccc}
9 & 5 & 6 & 7 \\
1 & 0 & 8 & 5 \\
\hline
1 & 0 & 6 & 5 & 2
\end{array}.$$

This process is fairly elaborate; in many cases one arrives at the solution a good deal more easily.

Problem Set 6.6

Try to analyze the following examples by the method we have just
illustrated.

1.   S END   4. A D A M
     M O R E   A N D
     G O L D   E V E
     ───────   O N
     MO NE Y   A
                          ─────────
                          R A F T

2.   H O C U S
      P O C U S   5.   S E E
     ─────────   S E E
     P R E S T O   S E E
                             Y E S
3. F O R T Y   ───────
        T E N   E A S Y .
        T E N
     ─────────

S I X T Y

If you want more, try to construct some of your own. If you are
familiar with computers, try to devise ways of programming the
solutions of such problems.

CHAPTER SEVEN

Congruences

7.1 Definition of Congruence

Number theory has an algebra of its own, known as the *theory of congruences*. Ordinary algebra originally developed as a shorthand for the operations of arithmetic. Similarly, congruences represent a symbolic language for divisibility, the basic concept of number theory. Gauss first introduced the notion of congruences.

Before we turn to congruences we shall make one remark about the numbers we shall study in this chapter. We began this book by saying that we were concerned with the positive integers 1, 2, 3, $\cdots$, and in previous chapters we have restricted ourselves to these and the additional number 0. But we have now reached a stage where it is advantageous to enlarge our scope to include all integers, positive or negative,

$$0, \pm 1, \pm 2, \pm 3, \cdots.$$

This does not in any essential way affect our previous concepts; in the following, when we talk about primes, divisors, greatest common divisors and the like, we shall continue to take them as positive integers.

Now let us turn to the language of congruences. If a and b are two integers and if their difference $a - b$ is divisible by m, we express this by writing

(7.1.1) $$a \equiv b \pmod{m}$$

and by saying

(7.1.1) a is congruent to b, modulo m.

83

The divisor m we suppose to be positive; it is called the *modulus* of the congruence. The statements (7.1.1) mean

(7.1.2) $a - b = mk,$ k an integer.

Examples:

1) $23 \equiv 8 \pmod{5}$ since $23 - 8 = 15 = 5 \cdot 3.$

2) $47 = 11 \pmod{9}$ since $47 - 11 = 36 = 9 \cdot 4.$

3) $-11 \equiv 5 \pmod{8}$ since $-11 - 5 = -16 = 8(-2).$

4) $81 \equiv 0 \pmod{27}$ since $81 - 0 = 81 = 27 \cdot 3.$

The last example shows that, in general, instead of saying that a number a is divisible by the number m we can write

$$a \equiv 0 \pmod{m},$$

for this means

$$a - 0 = a = mk,$$

where k is some integer. For instance, instead of saying that a is an even number we can write

$$a \equiv 0 \pmod{2}.$$

In the same manner one sees that an odd number is a number satisfying

$$a \equiv 1 \pmod{2}.$$

This somewhat "odd" terminology is quite common in mathematical papers.

7.2 Some Properties of Congruences

The way in which we write congruences reminds us of equations, and, indeed, congruences and algebraic equations have a number of properties in common. The simplest are the three following:

(7.2.1) $a \equiv a \pmod{m};$

this is a consequence of $a - a = 0 = m \cdot 0.$

(7.2.2) $a \equiv b \pmod{m}$ implies $b \equiv a \pmod{m}$.

This follows from $b - a = -(a - b) = m(-k)$.

(7.2.3) $a \equiv b \pmod{m}$, and $b \equiv c \pmod{m}$

imply

$$a \equiv c \pmod{m},$$

because the first two statements mean

$$a - b = mk, \qquad b - c = ml,$$

so that

$$a - c = (a - b) + (b - c) = m(k + l).$$

Example. From

$$13 \equiv 35 \pmod{11}, \qquad 35 \equiv -9 \pmod{11},$$

it follows that

$$13 \equiv -9 \pmod{11}.$$

We said that congruences resemble equalities in their properties. Actually we can consider equality to be a type of congruence, namely, congruence modulo 0. By definition

$$a \equiv b \pmod{0}$$

means

$$a - b = 0 \cdot k = 0 \qquad \text{or} \qquad a = b.$$

You will almost never encounter this congruence form of writing equations in the mathematical literature. But there is another congruence, apparently quite trivial, which is in use occasionally. When the modulus is $m = 1$, we have

(7.2.4) $a \equiv b \pmod{1}$

for any pair of integers a and b, for this just means that

(7.2.5) $a - b = 1 \cdot k = k$

is an integer. But let us now suppose, for a brief instant, that a and b are arbitrary real numbers, not necessarily integers. Then the fact that they are congruent (mod 1) means that their difference is integral, that is, the two numbers have the same fractional part (or decimal part if they are written with decimals).

Example.

$$8\tfrac{1}{3} \equiv 1\tfrac{1}{3} \pmod 1$$

or

$$8.333\cdots \equiv 1.333\cdots \pmod 1.$$

Let us return to the properties of ordinary congruences of integers; from now on we suppose always that the modulus is an integer $m > 2$.

We can divide the number axis, from the origin in both directions, into intervals of length m as in Figure 7.2.1. Then every integer a, positive or negative, falls into one of these intervals or on a dividing line, so we can write

$$(7.2.6) \qquad\qquad a = km + r,$$

where k is some integer and r one of the numbers

$$(7.2.7) \qquad\qquad 0, 1, 2, \cdots, m - 1.$$

This is a slight generalization of the division of positive integers in Section 4.3. Also, here we call r in (7.2.6) the *remainder* of a when it is divided by m, or remainder $\pmod m$.

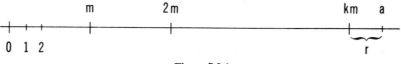

Figure 7.2.1

Examples:

1) $a = 11,$ $m = 7,$ $11 = 7 \cdot 1 + 4.$

2) $a = -11,$ $m = 7,$ $-11 = 7(-2) + 3.$

The division (7.2.6) can also be written as a congruence

$$(7.2.8) \qquad\qquad a \equiv r \pmod m,$$

so each number is congruent to its remainder $\pmod m$. In the examples above we have

$$11 \equiv 4 \pmod 7, \qquad -11 \equiv 3 \pmod 7.$$

No two of the remainders in (7.2.7) are congruent (mod m), for the difference between any two of them is less than m. Therefore, two numbers which are incongruent (not congruent) (mod m) must have different remainders. So we conclude:

A congruence $a \equiv b \pmod{m}$ holds if and only if a and b are numbers which have the same remainder when divided by m.

There is another way of conceiving of this congruence. For the moment, let a and b be positive integers. We observed in our discussion of number systems in Section 6.2 that when a is written to the base m,

$$a = (a_n, \cdots, a_1, a_0)_m,$$

the last digit a_0 is the remainder of a when it is divided by m. If we use this observation to reword our interpretation of a congruence, we may say:

The congruence $a \equiv b \pmod{m}$ holds for the (positive) integers a and b if and only if a and b have the same last digit to the base m.

For example,

$$37 \equiv 87 \pmod{10}$$

since the two numbers have the same last digit in the decimal number system.

Problem Set 7.2

1. Find the remainders $-37 \pmod{7}$, $-111 \pmod{11}$, $-365 \pmod{30}$.

7.3 The Algebra of Congruences

From algebra we recall that equations can be added, subtracted, multiplied. Exactly the same rules hold for congruences. Suppose we have the congruences

(7.3.1) $\qquad a \equiv b \pmod{m}, \qquad c \equiv d \pmod{m},$

By definition this means that

(7.3.2) $a = b + mk, \qquad c = d + ml,$

where k and l are integers. Let us add these equations (7.3.2). The result is

$$a + c = b + d + m(k + l)$$

which may be written

(7.3.3) $a + c \equiv b + d \pmod{m};$

in other words, two congruences can be added. In the same manner one shows that one congruence can be subtracted from another, that is,

(7.3.4) $a - c \equiv b - d \pmod{m}.$

Example.

(7.3.5) $11 \equiv -5 \pmod 8, \qquad$ and $\qquad 7 \equiv -9 \pmod 8.$

By adding one obtains

$$18 \equiv -14 \pmod 8,$$

and by subtracting,

$$4 \equiv 4 \pmod 8.$$

Both of these are true congruences.

One can also multiply two congruences. From (7.3.1) and (7.3.2) follows

$$ac = bd + m(kd + bl + mkl),$$

so that

(7.3.6) $ac \equiv bd \pmod m.$

Example. When the two congruences in (7.3.5) are multiplied one obtains

$$77 \equiv 45 \pmod 8.$$

A congruence

$$a \equiv b \pmod m$$

can be multiplied by any integer c to give

(7.3.7) $ac \equiv bc \pmod{m}$.

This one can consider a special case of the multiplication (7.3.6) for $c = d$; or it follows directly from the definition of congruence.

Example. When the first congruence in (7.3.5) is multiplied by 3 one finds

$$33 \equiv -15 \pmod{8}.$$

It is a natural question to ask when, in a congruence (7.3.7), one can cancel the common factor c and obtain a correct congruence

$$a \equiv b \pmod{m}.$$

At this point the congruences differ from equations. For instance, one has

$$22 \equiv -2 \pmod{8},$$

but the cancellation of the factor 2 would give

$$11 \equiv -1 \pmod{8}$$

which is not true.

There is one important case in which cancellation is permitted:

If $ac \equiv bc \pmod{m}$, then $a \equiv b \pmod{m}$ provided m and c are relatively prime.

PROOF: The first congruence means

$$ac - bc = (a - b)c = mk.$$

If $(m, c) = 1$, it follows that $a - b$ is divisible by m according to a result we proved in Section 4.2 (page 42).

Example. In the congruence

$$4 \equiv 4^8 \pmod{11}$$

we can cancel the factor 4, since $(11, 4) = 1$. This yields

$$1 \equiv 12 \pmod{11}.$$

Problem Set 7.3

Construct for yourself examples for the various congruence rules given above.

7.4 Powers of Congruences

Suppose again that we have a congruence

$$a \equiv b \pmod{m}.$$

As we just saw, we can multiply this congruence by itself to obtain

$$a^2 \equiv b^2 \pmod{m}.$$

In general, one can multiply the congruence by itself often enough to obtain

$$a^n \equiv b^n \pmod{m}$$

for any positive integer n.

Example. From

$$8 \equiv -3 \pmod{11}$$

follows, by squaring,

$$64 \equiv 9 \pmod{11},$$

and by taking the third power,

$$512 \equiv -27 \pmod{11}.$$

Many of the results on congruences are concerned with finding the remainders of high powers of a number. Let us indicate how one may proceed. Suppose, for instance, that we want to find the remainder of

$$3^{89} \pmod{7}.$$

One way of doing this is by repeated squarings. We find

$$9 = 3^2 \equiv 2 \pmod{7}$$
$$3^4 \equiv 4$$
$$3^8 \equiv 16 \equiv 2$$

$$3^{16} \equiv 4$$
$$3^{32} \equiv 16 \equiv 2$$
$$3^{64} \equiv 4 \pmod 7.$$

Since

$$89 = 64 + 16 + 8 + 1 = 2^6 + 2^4 + 2^3 + 1,$$

it follows that

$$3^{89} = 3^{64} \cdot 3^{16} \cdot 3^8 \cdot 3 \equiv 4 \cdot 4 \cdot 2 \cdot 3 \equiv 5 \pmod 7.$$

Thus the remainder (mod 7) is 5; in other words, it follows from what we said in Section 7.2 that in the system with base 7 the last digit of 3^{89} is 5.

Actually what we have done in order to find this remainder is to write the exponent

$$89 = 2^6 + 2^4 + 2^3 + 1 = (\,1, \ 0, \ 1, \ 1, \ 0, \ 0, \ 1\,)_2$$

in the binary number system. By repeated squarings we found the remainders (mod 7) of the various binary powers

$$1, \ 2, \ 4, \ 8, \ 16, \ 32, \ 64.$$

A corresponding method can always be used. But special cases can often be handled much more simply by astute observations. For instance, in the case above we note that

$$3^3 \equiv -1 \pmod 7,$$
$$3^6 \equiv 1 \pmod 7,$$

so we conclude that

$$3^{84} = (3^6)^{14} \equiv 1 \pmod 7.$$

Therefore

$$3^{89} = 3^{84} \cdot 3^3 \cdot 3^2 \equiv 1 \cdot (-1) \cdot 2 = -2 \equiv 5 \pmod 7$$

as before.

As another illustration we may consider the Fermat numbers which we introduced in Section 2.3:

$$F_t = 2^{2^t} + 1.$$

The first ones are

$$F_0 = 3, \quad F_1 = 5, \quad F_2 = 17, \quad F_3 = 257, \quad F_4 = 65537.$$

This seems to suggest:

The decimal numerals for all Fermat numbers except F_0 and F_1 end in the digit 7.

Let us prove by congruences that this is the case. Evidently it is the same as saying that the numbers

$$2^{2^t}, \quad t = 2, 3, \cdots$$

end in the digit 6. This we prove by induction. We observe that

$$2^{2^2} = 16 \qquad \equiv 6 \ (\text{mod } 10),$$

$$2^{2^3} = 256 \qquad \equiv 6 \ (\text{mod } 10),$$

$$2^{2^4} = 65536 \equiv 6 \ (\text{mod } 10).$$

Moreover, if we square 2^{2^k} the result is

$$(2^{2^k})^2 = 2^{2 \cdot 2^k} = 2^{2^{k+1}}.$$

Suppose that for some t

$$2^{2^t} \equiv 6 \ (\text{mod } 10);$$

by squaring this congruence we find

$$2^{2^{t+1}} \equiv 36 \equiv 6 \ (\text{mod } 10)$$

as desired.

7.5 Fermat's Congruence

From algebra we recall the binomial law

$$x + y = x + y$$

(7.5.1) $$(x + y)^2 = x^2 + 2xy + y^2$$

$$(x + y)^3 = x^3 + 3x^2y + 3xy^2 + y^3$$

$$(x + y)^4 = x^4 + 4x^3y + 6x^2y^2 + 4xy^3 + y^4,$$

and in general

$$(7.5.2) \quad (x+y)^p = x^p + \binom{p}{1} x^{p-1}y + \binom{p}{2} x^{p-2}y^2 + \cdots + y^p.$$

Here the first and last coefficients are unity. The middle binomial coefficients are

$$\binom{p}{1} = \frac{p}{1}, \quad \binom{p}{2} = \frac{p(p-1)}{1 \cdot 2},$$

(7.5.3)

$$\binom{p}{3} = \frac{p(p-1)(p-2)}{1 \cdot 2 \cdot 3}, \quad \cdots,$$

and in general

$$(7.5.4) \quad \binom{p}{r} = \frac{p(p-1)(p-2)\cdots(p-r+1)}{1 \cdot 2 \cdots r},$$

$$r = 1, 2, \cdots, p-1.$$

Since these coefficients are obtained by the successive multiplications by $x+y$ indicated in (7.5.1), it is clear that they are integers.

Suppose from now on that p is a prime. To write the integers (7.5.4) in integral form we must cancel all the common factors in the denominator

$$1 \cdot 2 \cdots r$$

and in the numerator

$$p(p-1)\cdots(p-r+1).$$

But the denominator does not include the prime factor p, so that after the cancellation, p is still present in the numerator. We conclude:

All the binomial coefficients (except the first and last) in (7.5.2) are divisible by p if p is a prime.

Now let x and y in (7.5.2) be integers. If we take the formula (7.5.2) as a congruence (mod p), we may conclude:

For integers x and y and any prime p,

$$(7.5.5) \quad (x+y)^p \equiv x^p + y^p \pmod{p}.$$

As an example let us take $p = 5$, so

$$(x + y)^5 = x^5 + 5x^4y + 10x^3y^2 + 10x^2y^3 + 5xy^4 + y^5.$$

Since all middle coefficients are divisible by 5, we find

$$(x + y)^5 \equiv x^5 + y^5 \ (\text{mod } 5)$$

corresponding to (7.5.5).

We can deduce some important consequences from the congruence (7.5.5). Let us apply it first to the case $x = y = 1$. It gives us

$$2^p = (1 + 1)^p \equiv 1^p + 1^p = 2 \ (\text{mod } p).$$

In the next step we take $x = 2$, $y = 1$ and find

$$3^p = (2 + 1)^p \equiv 2^p + 1^p;$$

then we use the previous result $2^p \equiv 2 \ (\text{mod } p)$ to obtain

$$2^p + 1^p \equiv 2 + 1 \equiv 3 \ (\text{mod } p), \quad \text{so} \quad 3^p \equiv 3 \ (\text{mod } p).$$

Next, for $x = 3$, $y = 1$, we get

$$4^p \equiv 4 \ (\text{mod } p).$$

Using this process, we can prove by induction that $a^p \equiv a \ (\text{mod } p)$ for all values

(7.5.6) $$a = 0, 1, \cdots, p - 1;$$

the special cases $a = 0$ and $a = 1$ are self-evident. Since every number is congruent (mod p) to one of the remainders in (7.5.6), we conclude:

For any integer a and any prime p,

(7.5.7) $$a^p \equiv a \ (\text{mod } p).$$

This congruence law is commonly called Fermat's theorem, although some writers call it the *little* Fermat theorem to distinguish it from Fermat's last theorem or Fermat's conjecture which we mentioned in Section 5.3.

Example. For $p = 13$ and $a = 2$ we find $13 = 8 + 4 + 1$, so $2^{13} = 2^{8+4+1} = 2^8 \cdot 2^4 \cdot 2^1$. Since

$$2^4 = 16 \equiv 3 \ (\text{mod } 13), \quad 2^8 \equiv 9 \ (\text{mod } 13),$$

we obtain

$$2^{13} = 2^8 \cdot 2^4 \cdot 2 \equiv 9 \cdot 3 \cdot 2 \equiv 2 \pmod{13}$$

as stated by Fermat's congruence.

According to the cancellation law for congruences stated at the end of Section 7.3, we can cancel the common factor a on both sides in Fermat's congruence (7.5.7), provided that a is relatively prime to the modulus p. This gives us the result:

If a is an integer not divisible by the prime p, then

(7.5.8) $a^{p-1} \equiv 1 \pmod{p}$.

This result is also known as *Fermat's theorem*.

Example. When $a = 7$, $p = 19$, we find

$$7^2 = 49 \equiv 11 \pmod{19}$$

$$7^4 \equiv 121 \equiv 7 \pmod{19}$$

$$7^8 \equiv 49 \equiv 11 \pmod{19}$$

$$7^{16} \equiv 121 \equiv 7 \pmod{19},$$

and this gives

$$a^{p-1} = 7^{18} = 7^{16} \cdot 7^2 \equiv 7 \cdot 11 \equiv 1 \pmod{19}$$

as required by Fermat's congruence (7.5.8)

As an application of Fermat's congruence (7.5.8) we return to the Pythagorean triangles discussed in Chapter 5 and we prove:

The product of the sides of a Pythagorean triangle is divisible by 60

PROOF: Evidently it suffices to prove this for primitive triangles. According to the formulas (5.2.7) the product is

$$P = 2mn(m^2 - n^2)(m^2 + n^2) = 2mn(m^4 - n^4).$$

The number P is divisible by 60 if and only if it is divisible by 4, by 3 and by 5. Since one of the numbers m and n is even, $2mn$ and hence also P is divisible by 4. It is divisible by 3 when at least one of the numbers m and n is divisible by 3; but also if neither m nor n is divisible by 3, P is because, by (7.5.8), $(m, 3) = 1$ and $(n, 3) = 1$ imply $m^2 \equiv 1 \pmod 3$ and $n^2 \equiv 1 \pmod 3$, so that

$$m^2 - n^2 \equiv 1 - 1 \equiv 0 \pmod 3.$$

Similarly, P is divisible by 5. This is evident if m or n is divisible by 5. If neither of them is divisible by 5 we have, again by Fermat's congruence (7.5.8),

$$m^4 - n^4 \equiv 1 - 1 \equiv 0 \pmod 5.$$

CHAPTER EIGHT

Some Applications of Congruences

8.1 Checks on Computations

As we have mentioned, the creator of congruence theory was the German mathematician Gauss. His famous work on number theory, the *Disquisitiones Arithmeticae*, appeared in 1801 when he was twenty-four years old. The *Disquisitiones* has recently been translated into English (Yale University Press), so if you are interested in reading parts of one of the masterpieces of mathematics you may now do so. The first chapters deal with congruence theory and you have already learned enough about congruences to be able to follow Gauss's presentation.

But let us not fail to mention that there are traces of congruence theory centuries before the time of Gauss. Some of these appear in the ancient check rules for arithmetical computations. They formed an integral part of the instruction in arithmetic in the Renaissance. Some of them are still in use, and for all we know about their origin they may have their roots in antiquity.

How they originally were introduced we don't know, but let us indicate a plausible way in which they may have been discovered. We go back to the times of the computing boards. On such an abacus each digit in the numbers needed for the calculations would be laid out by means of counters or stones or sticks or nuts, each group marking the number of units, tens, hundreds, and so on, according to its place. A number in our decimal system

$$(8.1.1) \quad N = a_n \cdot 10^n + a_{n-1} \cdot 10^{n-1} + \cdots + a_2 \cdot 10^2 + a_1 \cdot 10 + a_0$$

$$= (a_n, \ a_{n-1}, \ \cdots, \ a_2, \ a_1, \ a_0)_{10}$$

would require a total of

$$(8.1.2) \quad S_N = a_n + a_{n-1} + \cdots + a_2 + a_1 + a_0$$

counters. This number we call the *digit sum* of N.

Suppose now that we wish to perform a simple operation on the board, namely: to add two numbers N and M. We then mark also the second number

$$M = (b_m, \ b_{m-1}, \ \cdots, \ b_2, \ b_1, \ b_0)_{10}$$

on the board by means of

$$S_M = b_m + b_{m-1} + \cdots + b_2 + b_1 + b_0$$

additional counters on the same lines. On some lines there may now be more than nine counters. The operation required to find $N + M$ consists in replacing ten counters on one line by a single counter on the next line and continuing until no further such reductions can be made. In each step one replaces ten counters by a single one, so there is a net loss of nine counters on the board. Thus we see that if the addition is performed correctly the number of counters remaining on the board must satisfy

$$(8.1.3) \quad S_{N+M} \equiv S_N + S_M \ (\mathrm{mod}\ 9),$$

that is, the number of counters still on the board must differ from the original total by a multiple of 9. This check (8.1.3) still carries its old name: casting out nines.

After this rule had been discovered it cannot have taken long to notice that it applies also to several summands, to a difference, and to a product; in the last case we have, analogous to (8.1.3),

$$(8.1.4) \quad S_M \cdot S_N \equiv S_{MN} \ (\mathrm{mod}\ 9).$$

To prove these rules theoretically is an easy task when we use congruences. It is evident that

$$(8.1.5) \quad 1 \equiv 1, \quad 10 \equiv 1, \quad 10^2 \equiv 1, \quad 10^3 \equiv 1, \quad \cdots \ (\mathrm{mod}\ 9),$$

so from (8.1.1) and (8.1.2) we conclude that

(8.1.6) $N \equiv S_N \pmod 9$.

Therefore, from the congruence rules which we established in Section 7.3, it is clear that

$$S_N \pm S_M \equiv N \pm M \equiv S_{N \pm M}, \quad S_N \cdot S_M \equiv N \cdot M \equiv S_{N \cdot M} \pmod 9.$$

The casting out of nines is mostly applied to multiplications. Take as an example the numbers

(8.1.7) $M = 3119, \quad N = 3724$

and the product

$$M \cdot N = 11614156.$$

This calculation cannot be correct, for if it were we would have

$$M \equiv S_M \equiv 3 + 1 + 1 + 9 \equiv 5 \pmod 9,$$
$$N \equiv S_N \equiv 3 + 7 + 2 + 4 \equiv 7 \pmod 9,$$

and

$$MN \equiv S_{MN} \equiv 1 + 1 + 6 + 1 + 4 + 1 + 5 + 6 \equiv 7 \pmod 9.$$

But

$$5 \cdot 7 = 35 \equiv 8 \not\equiv 7 \pmod 9.$$

Actually the product should be

$$M \cdot N = 11615156.$$

In the medieval schools the pupils were strictly instructed to include the checks in their exercises. So in the manuscripts from these times one finds an added set of cross-bones which in our example (8.1.7) would appear as follows:

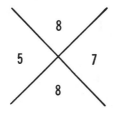

Figure 8.1.1

Here the numbers 5 and 7 in the side spaces represent the remainders of M and N (mod 9) and the upper 8 is the remainder of the calculated product $M \cdot N$. This should check with the product of the remainders in the lower space, here

$$5 \cdot 7 = 35 \equiv 8 \ (\text{mod } 9).$$

These cross-bone checks appear quite commonly in early printed arithmetic texts, for instance, in English texts from the seventeenth and eighteenth centuries. It is, of course, possible that a calculation contains an error not discerned by the method of casting out nines; but then we know that the mistake is an "error modulo 9".

It is clear that for other bases a similar check could be used. For a number

$$M = m_n b^n + m_{n-1} b^{n-1} + \cdots + m_2 b^2 + m_1 b + m_0$$

to the base b one has, as in (8.1.5),

$$1 \equiv 1, \quad b \equiv 1, \quad b^2 \equiv 1, \quad \cdots \ (\text{mod } b-1);$$

so, as before,

$$M \equiv S_M = m_n + m_{n-1} + \cdots + m_2 + m_1 + m_0 \ (\text{mod } b-1),$$

and the checking rules are the same.

This seemingly quite trivial observation has applications even in our ordinary decimal system. We mentioned in Section 7.5 that if we divide the digits of a decimal number into groups of three, then this grouping can be conceived of as an expansion of the number to the base

$$b = 10^3 = 1000.$$

Similarly, if one groups the digits in pairs, this corresponds to an expansion with base

$$b = 10^2 = 100.$$

Taking the numbers 3119 and 3724 as an example again, and writing

$$M = 31 \ 19, \qquad N = 37 \ 24$$

$$M \cdot N = 11 \quad 61 \quad 51 \quad 56,$$

we find

$$M \equiv 31 + 19 = 50 \ (\text{mod } 99), \qquad N \equiv 37 + 24 = 61 \ (\text{mod } 99),$$

$$M \cdot N = 11 + 61 + 51 + 56 = 179 \equiv 80 \ (\text{mod } 99).$$

Rechnung auff
der Linien vnnd Federn /
Auff allerley Handtierung /
Gemacht durch
Adam Riesen.

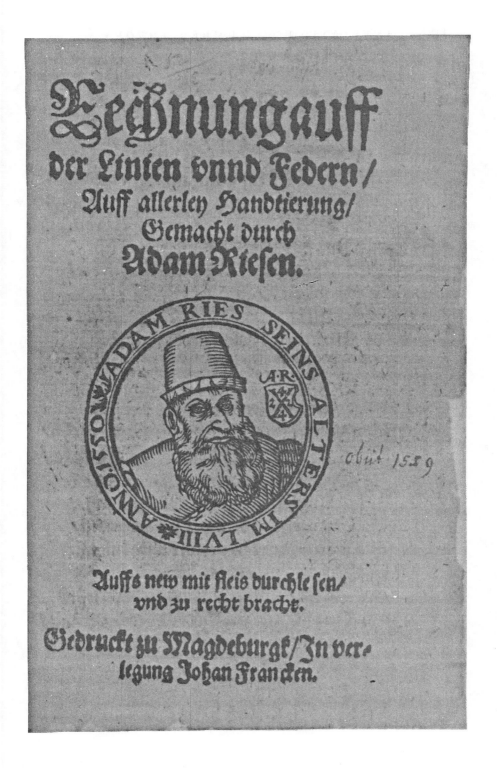

obiit 1559

Auffs new mit fleis durchlesen /
vnd zu recht bracht.

Gedruckt zu Magdeburgk/In ver-
legung Johan Francken.

Here our cross-bone check is

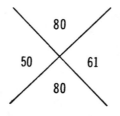

because, as one sees,

$$50 \cdot 61 \equiv 80 \ (\text{mod } 99).$$

This check is more efficient than casting out nines because the modulus is a larger one, and so the probability that the answer is right is correspondingly greater. In other words, an "error modulo 99" is less likely than an "error modulo 9".

8.2 The Days of the Week

Many problems of astronomy and chronology involving periodicity can be formulated in terms of number theoretical concepts. We shall give a single example: The determination of the day of the week on which a given date will fall. The week days repeat themselves in periods of 7 so instead of the usual names we can give each day a number:

Sunday $= 0$,   M $= 1$,   Tu $= 2$,   W $= 3$,   Th $= 4$,   F $= 5$,   S $= 6$.

If we do this, each integer corresponds to a week-day, namely, the day determined by its remainder (mod 7).

If we had the pleasant situation that the number of days in a year was divisible by 7, all dates would fall on the same week-day every year, so schedule making would be simple and calendar printers would have a much reduced business. However, the number of days in the year is

$$365 \equiv 1 \ (\text{mod } 7),$$

except for leap years when it is

$$366 \equiv 2 \ (\text{mod } 7).$$

This shows that for an ordinary year the week-day number W of a given date will increase by 1 in the next year; for instance, if January 1 is on a Sunday in one year, it will fall on a Monday the next. This is not too complicated, but the simple pattern is broken by the leap years. This occurs every fourth year, and then the week-day number increases by 2; moreover, we have the additional difficulty that the leap year day is not added at the beginning or end of the year, but on February 29, within the year. This makes it convenient in the general formula for W, which we shall give below, to agree to *count March as the first month, April as the second, and so on, until January is the eleventh month and February the twelfth month in the preceding year.*

But our troubles are not over. In the Julian calendar, introduced by order of Julius Caesar, the year was taken to be exactly $365\frac{1}{4}$ days corresponding to the leap year rule. This, however, is not quite correct since the astronomical year is actually

$$365.2422 \text{ days.}$$

This small error caused a gradual change of the seasons in relation to the calendar; for instance, in the sixteenth century the vernal equinox (first day of Spring) fell on March 11 instead of March 21, as it was originally supposed to.

To remedy this situation Pope Gregory XIII, after much hesitation, introduced his calendar reform for the Catholic countries in the year 1582. Ten days were omitted from this year by making Friday, October 5, into Friday, October 15. Furthermore, to keep the calendar in step the following Gregorian rules for the leap years were introduced.

The century years

$$1700, \quad 1800, \quad 1900, \quad 2100, \quad 2200, \quad 2300, \quad \cdots,$$

where the number of centuries is not divisible by 4, are not leap years. The remaining century years

$$1600, \quad 2000, \quad 2400, \quad \cdots$$

continue to be leap years. One obtains a very close approximation to the correct length of the year, but now it is a trifle too short. It has been proposed to drop the years 4000, 8000, $\cdots$ as leap years contrary to the Gregorian rule; since the question is still open and of no concern to the near future we disregard it in our formula

Suppose now that we have a given date: The d-th day in the m-th month, counting m as explained above, in the year

(8.2.1) $N = C \cdot 100 + Y$,

where C is the number of centuries and Y the year number within the century. Then one can prove that our week-day number is determined by the congruence

(8.2.2)

$$W \equiv d + [\tfrac{1}{5}(13m - 1)] + Y + [\tfrac{1}{4}Y] + [\tfrac{1}{4}C] - 2C \pmod{7}.$$

We recall from Section 4.3 that the brackets which occur in the formula denote the greatest integer contained in the number.

Example. Pearl Harbor day, December 7, 1941. Here

$$d = 7, \quad m = 10, \quad C = 19, \quad Y = 41$$

so that

$$W \equiv 7 + 25 + 41 + 10 + 4 - 38 \equiv 0 \pmod{7};$$

that is, it was on a Sunday.

Example. On what day will January 1, 2000, fall? Here

$$d = 1, \quad m = 11, \quad C = 19, \quad Y = 99$$

and

$$W \equiv 1 + 28 + 1 + 3 + 4 - 38 \equiv 6 \pmod{7},$$

so the first day in the next century will be a Saturday.

In connection with such calculations as these it must be noted that the formula cannot be applied before the Gregorian calendar was introduced. In England and English colonies this occurred in 1752, when eleven days were dropped by changing September 3 into September 14, New Style.

You may wish to understand in detail how the formula (8.2.2) has been established. If not, skip the rest of this section. We divide the analysis into two parts.

First, let us determine the week-day number of March 1 in any year N in (8.2.1). We take some arbitrary starting year, say 1600, and call its week-day number d_{1600}. We could look up old records and find out what it was, but this is not necessary; it will come out as a consequence of our considerations.

If there were no leap years we would find the week-day d_N of March 1 in the year N simply by adding one day to d_{1600} for each year which has passed. This gives us the number

(8.2.3) $$d_{1600} + (100C + Y - 1600) \pmod 7.$$

Taking the leap years into account and assuming that they follow regularly every fourth year, we should add to this

(8.2.4) $$[\tfrac{1}{4}(100C + Y - 1600)] = 25C - 400 + [\tfrac{1}{4}Y].$$

This is a little too much since the even centuries ordinarily are not leap years so we should subtract

(8.2.5) $$C - 16$$

days on this account. But we had the exception that when the century number C is divisible by 4 the year $100C$ is still a leap year, so we should add a final correction

(8.2.6) $$[\tfrac{1}{4}(C - 16)] = [\tfrac{1}{4}C] - 4.$$

We now add the expressions (8.2.3) and (8.2.4), subtract (8.2.5) and add (8.2.6). This gives us as the week-day for March 1 in the year N

$$d_N \equiv d_{1600} + 124C + Y - 1988 + [\tfrac{1}{4}C] + [\tfrac{1}{4}Y] \pmod 7.$$

To simplify we reduce the numbers $\pmod 7$ and this leads to

(8.2.7) $$d_N \equiv d_{1600} - 2C + Y + [\tfrac{1}{4}C] + [\tfrac{1}{4}Y] \pmod 7.$$

Let us apply this formula to the year $N = 1968$ in which March 1 falls on a Friday; hence $d_{1968} = 5$. Here

$$C = 19, \quad [\tfrac{1}{4}C] = 4, \quad Y = 68, \quad [\tfrac{1}{4}Y] = 17,$$

and we find

$$d_{1968} = 5 \equiv d_{1600} + 2 \pmod 7.$$

This gives us $d_{1600} = 3$, so March 1, 1600, fell on a Wednesday. When this is plugged in (8.2.7) we arrive at the formula

(8.2.8) $$d_N \equiv 3 - 2C + Y + [\tfrac{1}{4}C] + [\tfrac{1}{4}Y] \pmod 7$$

for the week-day of March 1 in any year N.

Secondly, we shall have to determine the number of days from

March 1 to any other day in the year (mod 7). Since the number of days in the months varies, it requires a little trick to express the additions mathematically. Let us begin by finding the number of days to be added to the day number of March 1 to obtain the day number of any other 1st in a month.

Since March has 31 days one should add 3 to get the number of April 1st; to get May 1st we must add $3 + 2$ days, since April has 30 days. By proceeding successively, one arrives at the following table of additions:

I	March	0	VII	September	16
II	April	3	VIII	October	18
III	May	5	IX	November	21
IV	June	8	X	December	23
V	July	10	XI	January	26
VI	August	13	XII	February	29

It is worth noting that by beginning our count of the year from March 1 we have actually returned to the ancient Roman calendar introduced by Julius Caesar, with September, October, November, December as the seventh, eighth, ninth, tenth months, as their Latin names state.

Let us go back to our list of additions. The numbers in the tables, although irregular, proceed at an average increase of

$$\tfrac{29}{11} = 2.6\cdots \text{ pr. month.}$$

Since the first term is 0 we should subtract about 2.6 and take the next integer below. This turns out to be not entirely correct, but by juggling the subtracted term we arrive at the expression

$$(8.2.9) \quad [2.6m - 2.2] = [\tfrac{1}{5}(13m - 11)], \quad m = 1, 2, \cdots, 12.$$

Marvelous to say, all is now well: If you check the values for $m = 1, \cdots, 12$ in (8.2.9) we get exactly the values in our table.

Therefore, the expression (8.2.9) should be added to the March 1 number (8.2.8) to obtain the week-day of the first day in the m-th month. Finally, since we want the number of the d-th day of this month we should add $d - 1$, and when this is done and a slight rearrangement of the terms is performed, we arrive exactly at our stated formula (8.2.2).

Problem Set 8.2

1. Find the week-day of your own birth.

2. How does the formula (8.2.2) simplify when one considers only the years 1900–1999?

3. How do the week-days of the date of birth distribute in your class?

8.3 Tournament Schedules

As another simple application of congruence theory we may take the preparation of round-robin schedules such as are used in all sorts of competitions from chess to baseball.

We assume that there are N participants or teams. When N is an odd number, we cannot pair all teams in each round; there will always be one which has a bye. We can eliminate this difficulty by adding a fictitious team T_0 and making up a schedule for $N + 1$ teams, including T_0. In each round the team scheduled to play T_0 is given this round off.

We may suppose, therefore, that there is an even number of teams N. We give each of them a number

$$x = 1, 2, \cdots, N - 1, N.$$

The total number of rounds played by each team is $N - 1$.

Suppose now that x belongs to the set

$$(8.3.1) \qquad 1, 2, \cdots, N - 1.$$

As opponent to x in the r-th round we assign the team y_r from the set (8.3.1), where y_r is the number determined by the congruence

$$(8.3.2) \qquad x + y_r \equiv r \pmod{N-1}.$$

To see that different x's will have different opponents y_r, note that

$$x + y_r \equiv r \equiv x' + y_r \pmod{N-1}$$

implies

$$x \equiv x' \pmod{N-1},$$

or $x = x'$ since these numbers belong to (8.3.1).

The only complication arises in the case where $x = y_r$, and so in (8.3.2)

(8.3.3) $$2x \equiv r \pmod{N-1}.$$

There is only a single x in (8.3.1) for which this can occur, for if

$$2x \equiv r \equiv 2x' \pmod{N-1},$$

it follows that

$$2(x - x') \equiv 0 \pmod{N-1},$$

or

$$x \equiv x' \pmod{N-1}$$

since $N - 1$ is odd. There is always a solution to (8.3.3) in (8.3.1), namely

$$x = \frac{r}{2} \qquad \text{when } r \text{ is even,}$$

$$x = \frac{r + N - 1}{2} \qquad \text{when } r \text{ is odd.}$$

Through the relation (8.3.2) we have assigned an opponent in the r-th round to each x in (8.3.1) with the exception of that x_0 which satisfies (8.3.3); this x_0 we match with the N-th team.

It remains to show that by these matchings each team plays a different opponent in each round $r = 1, \cdots, N - 1$. We verify it first for the somewhat exceptional N-th team. In the r-th round it plays the x_0 which is determined by (8.3.3). Suppose $s \neq r$; then in the s-th round N plays the team x_0' which satisfies

$$2x_0' \equiv s \pmod{N-1}.$$

One cannot have $x_0 = x_0'$ for it would lead to

$$2x_0 = 2x_0' \equiv r \equiv s \pmod{N-1},$$

and hence to $r = s$.

Consider next the various opponents of a team x in (8.3.1). It will play the N-th team just once, namely, for the r_0 defined by

$$2x \equiv r_0 \pmod{N-1}.$$

Suppose now that $r \neq r_0$ and $s \neq r_0$. Then the opponents of x in the r-th and s-th rounds will be determined by (8.3.2):

$$x + y_r \equiv r \pmod{N-1} \quad \text{and} \quad x + y_s \equiv s \pmod{N-1}.$$

Again, $y_r = y_s$ would lead to $r = s$ as before, so we conclude that $y_r \neq y_s$.

Let us construct a round-robin table for $N = 6$ players by means of the method given above. Some simple calculations give the result tabulated below. The entry in the r-th row, x-th column gives the opponent of player x in the r-th round.

r \ x	1	2	3	4	5	6
1	5	4	6	2	1	3
2	6	5	4	3	2	1
3	2	1	5	6	3	4
4	3	6	1	5	4	2
5	4	3	2	1	6	5

Problem Set 8.3

1. Construct a table for $N = 8$ players.

2. Show that, when $r = 2$, the teams 1, 2, $\cdots$, N are matched with N, $N - 1$, $\cdots$, 2, 1, respectively.

3. Why does team $N - 1$ always play team r in the r-th round except when $r = N - 1$? In the exceptional case which team does it play?

4. Verify that if x plays y in the r-th round then also by the formulas y plays x in this round.

8.4 Prime or Composite?

As a final application of congruences we shall discuss a method for examining whether a large number is prime or composite. It is a very efficient method, the best general method we have when it comes to the investigation of some particular number chosen at random. It is based on the congruence (7.5.8) of Fermat (page 95).

Let N be the number we wish to examine. We select a as some small number relatively prime to N. Very often it is convenient to take a as some small prime which does not divide N, for instance, $a = 2$, or $a = 3$, or $a = 5$. If N were a prime it would obey

$$(8.4.1) \qquad a^{N-1} \equiv 1 \ (\mathrm{mod} \ N)$$

according to Fermat's congruence. Consequently, if we check this congruence (8.4.1) and find that it does not hold, we know that N is composite.

Example. Take $N = 91$, and choose $a = 2$. Then

$$a^{N-1} = 2^{90} = 2^{64} \cdot 2^{16} \cdot 2^8 \cdot 2^2.$$

Moreover,

$$2^8 = 256 \equiv -17 \ (\mathrm{mod} \ 91),$$

$$2^{16} = (2^8)^2 \equiv (-17)^2 = 289 \equiv 16 \ (\mathrm{mod} \ 91),$$

$$2^{32} = (2^{16})^2 \equiv (16)^2 = 256 \equiv -17 \ (\mathrm{mod} \ 91),$$

$$2^{64} = (2^{32})^2 \equiv (-17)^2 = 289 \equiv 16 \ (\mathrm{mod} \ 91),$$

so that

$$2^{90} = 2^{64} \cdot 2^{16} \cdot 2^8 \cdot 2^2$$

$$\equiv 16 \cdot 16 \cdot (-17) \cdot 4 \equiv 64 \not\equiv 1 \ (\mathrm{mod} \ 91).$$

We conclude that N is composite. Actually, $91 = 7 \cdot 13$.

Our example is much too simple to show the real power of the method. By suitable programming for computers it is possible in this manner to establish that certain very large numbers are composite. Unfortunately, the method gives no indication of what the factors are; hence, in many instances we know that a number is not a prime, yet we have no inkling of what its factors may be.

This applies particularly to the Fermat numbers

$$F_n = 2^{2^n} + 1$$

which we discussed in Section 2.3. They are primes for $n = 0, 1,$ 2, 3, 4, as we noted. To test the number

$$F_5 = 2^{2^5} + 1 = 2^{32} + 1 = 4\,294\,967\,297$$

by Fermat's congruence, we can take $a = 3$. If F_5 were prime one should have

(8.4.2) $3^{2^{32}} \equiv 1 \pmod{F_5}$.

To compute the remainder of the power on the left one must square 32 times and in each step reduce the result $(\bmod F_5)$. We shall spare the reader the details. One finds that the congruence (8.4.2) does not hold; hence, F_5 is composite. The known factor 641 has been found by trial. This same method has been used to show that several higher Fermat numbers are not primes. For some of them we know factors, for others we do not.

If the congruence (8.4.1) holds for some a relatively prime to N, the number N may or may not be a prime. The cases where it holds for a composite number N are exceptions, so one would usually guess that N is a prime. However, for most purposes one would like to know definitely. This can be accomplished by refining the method, basing it upon the remark that N *is a prime in case* (8.4.1) *holds for the exponent* $N - 1$ *but not for any proper divisor of* $N - 1$.

There is another approach, effective for numbers N that are not too large. We take $a = 2$. Poulet and Lehmer have computed all values of $N \leqslant 100\,000$ which are exceptional in the sense that

(8.4.3) $2^{N-1} \equiv 1 \pmod{N}$,

yet N is composite; these numbers N are sometimes called *pseudoprimes*. For each of these numbers N they have also given the largest prime factors.

By means of Poulet's and Lehmer's tables one can determine the primality of any number $N \leqslant 100\,000\,000$ as follows: Check first whether the congruence (8.4.3) holds. If it does not, N is composite. If the congruence does hold and N is in the tables, it is also composite, and we can read off a prime factor from the tables. Finally, if (8.4.3) holds and N is not in the tables, it is a prime.

The smallest composite number N satisfying (8.4.3) is

$$N = 341 = 11 \cdot 31.$$

Below 1000 there are two others, namely

$$N = 561 = 3 \cdot 11 \cdot 17,$$

$$N = 645 = 3 \cdot 5 \cdot 43.$$

The number 561 is remarkable, for here the congruence (8.4.1) holds for *every* integer a relatively prime to N. Such peculiar numbers, we say, have the *Fermat property*; a lot of research has been done on such numbers. For literature and tables for these numbers see D. H. Lehmer: *Guide to Tables in the Theory of Numbers*.

Solutions to Selected Problems

Problem Set 1.3

For both problems consult Table 3, page 53.

Problem Set 1.4

1. Suppose we know that

$$T_{n-1} = \tfrac{1}{2}(n-1)n.$$

You may check that this is true for $n = 2, 3, 4$. From Fig. 1.4.3 one sees that T_n is obtained from T_{n-1} by adding n so that

$$T_n = T_{n-1} + n = \tfrac{1}{2}n(n+1).$$

2. From Fig. 1.4.4 one sees that to obtain P_n one must add to P_{n-1}

$$1 + 3(n-1) = 3n - 2.$$

If we know already that

$$P_{n-1} = \tfrac{1}{2}(3(n-1)^2 - (n-1))$$

(this is true for $n = 2, 3, 4$ according to the list (1.4.3)), then it follows that

$$P_n = P_{n-1} + 3n - 2 = \tfrac{1}{2}(3n^2 - n).$$

3. The n-th k-gonal number is obtained from the $(n-1)$-st by adding

$$(k-2)(n-1) + 1$$

113

and one derives the formula in the same way as in Problem **2**. Problems 2 and 3 could have been solved differently by dividing the points into triangles as indicated in Fig. 1.4.4 and using the formula for T_n Carry through the details of such a proof.

Problem Set 1.5

1. For instance

16	3	2	13
9	6	7	12
5	10	11	8
4	15	14	1

obtained by interchanging second and third lines in Dürer's square is also magic. Less trivial is

16	4	1	13
9	5	8	12
6	10	11	7
3	15	14	2

2. Since the numbers in a 4 x 4 magic square do not exceed 16, only two years, namely 1515 and 1516, are possible. The first is evidently excluded, and the second turns out to be impossible in constructing a square.

Problem Set 2.1

2. 1979.

3. The numbers 114 to 126 are all composite.

Problem Set 2.3

1. $n = 3, 5, 15, 17, 51, 85$.

2. One has

$$\frac{360°}{51} = 6 \cdot \frac{360°}{17} - \frac{360°}{3} .$$

3. The products of one to five Fermat primes give altogether

$$5 + 10 + 10 + 5 + 1 = 31$$

different numbers for which the polygon can be constructed. The largest value is

$$n = 3 \cdot 5 \cdot 17 \cdot 257 \cdot 65\ 537 = 4\ 294\ 967\ 295.$$

Problem Set 2.4

1. In each of the first ten hundreds there are respectively

$$24, \quad 20, \quad 16, \quad 16, \quad 17, \quad 14, \quad 16, \quad 14, \quad 15, \quad 14$$

primes.

2. There are 11 such primes.

Problem Set 3.1

1. $120 = 2^3 \cdot 3 \cdot 5; \quad 365 = 5 \cdot 73; \quad 1970 = 2 \cdot 5 \cdot 197.$

3. $360 = 2 \cdot 2 \cdot 90 = 2 \cdot 6 \cdot 30 = 2 \cdot 10 \cdot 18 = 6 \cdot 6 \cdot 10.$

4. A number is an even-prime only when it has the form $2k$ where k is odd. Suppose that a number n has an even-prime factorization

$$n = (2k_1) \cdot (2k_2) \cdots,$$

where there are at least 2 factors. This leads to another such factorization

$$n = (2k_1 \cdot k_2) \cdot 2 \cdots.$$

This is different from the first except when $k_2 = 1$. The same reasoning applies to the following $k_3, k_4 \cdots$. We conclude that if there is a unique even-prime factorization one must have

$$n = (2k) \cdot 2 \cdot 2 \cdots = k \cdot 2^\alpha, \quad k \text{ odd}.$$

This, as one readily sees, is a unique factorization where $k = 1$, that is, $n = 2^\alpha$. If $k > 1$ there is a further condition: k must be an ordinary prime. If $k = a \cdot b$ there is another factorization, namely

$$n = (2a) \cdot (2b) \cdot 2 \cdot 2 \cdots.$$

Problem Set 3.2

1. A prime has 2 divisors; a prime power p^α has $\alpha + 1$ divisors.

2. $\tau(60) = 12$, $\tau(366) = 8$, $\tau(1970) = 8$.

3. The largest number of divisors for numbers ≤ 100 is 12 and this is obtained for the numbers

$$72; \quad 84; \quad 90; \quad 96.$$

Problem Set 3.3

1. 24; 48; 60; 10 080.

2. 2^{13}; 180; 45 360.

3. 24 and 36.

4. Let the number of divisors be $r \cdot s$, r and s primes. Then

$$n = p^{rs-1} \quad \text{or} \quad n = p^{r-1} \cdot q^{s-1},$$

p, q primes.

Problem Set 3.4

1. 8128 and 33 550 336.

Problem Set 4.1

1. (a) $(360, 1970) = 10$;   (b) $(30, 365) = 5$.

2. Suppose $\sqrt{2}$ is rat onal

$$\sqrt{2} = \frac{a}{b}.$$

After cancellation one can assume that a and b have no common factor. By squaring one obtains

$$2b^2 = a^2.$$

Because of the unique factorization theorem a is divisible by 2, consequently a^2 is divisible by 4. Again, by the uniqueness of prime factorization b^2 and so also b is divisible by 2, contrary to our assumption that a and b have no common factor. This contradiction shows that our initial rational expression for $\sqrt{2}$ cannot exist.

Problem Set 4.2

1. The odd numbers.

2. If p were any prime dividing n and $n + 1$ it would also have to divide

$$(n + 1) - n = 1.$$

3. None of them are relatively prime.

4. Yes.

Problem Set 4.3

2. $(220, 284) = 4$, $(1184, 1210) = 2$,

$(2620, 2924) = 4$, $(5020, 5564) = 4$.

3. To determine the highest power of 10 which divides

$$n! = 1 \cdot 2 \cdot 3 \cdot \, \cdots \, \cdot n$$

we first find the highest power of 5 which divides it. Every fifth number

$$5, \quad 10, \quad 15, \quad 20, \quad 25, \quad 30$$

is divisible by 5 and altogether there are $\left[\frac{n}{5}\right]$ of these up to n. But some of these are divisible by the second power of 5, namely 25, 50, 75, 100, $\cdots$, and there are $\left[\frac{n}{25}\right]$ of these. The ones which are divisible also by the third power $125 = 5^3$ are $125, 250, 275$, and of these there are $\left[\frac{n}{5^3}\right]$, etc. This shows that the exponent of the exact power of 5 which divides $n!$ is

(E) $$\left[\frac{n}{5}\right] + \left[\frac{n}{5^2}\right] + \left[\frac{n}{5^3}\right] + \cdots$$

where one continues the terms until the denominator exceeds the numerator.

Exactly the same argument applies to finding the power of any other prime p. In particular, when $p = 2$ one finds the exponent

$$\left[\frac{n}{2}\right] + \left[\frac{n}{2^2}\right] + \left[\frac{n}{2^3}\right].$$

Clearly this exponent is not less than the expression (E), so in $n!$ each factor 5 can be combined with a factor 2. Thus (E) also gives the exact power of 10 dividing $n!$, that is, the number of final zeros.

Examples: $n = 10$, $\left[\frac{10}{5}\right] = 2$, $\left[\frac{10}{5^2}\right] = 0$, so 10! ends in 2 zeros.

$n = 31$, $\left[\frac{31}{5}\right] = 6$, $\left[\frac{31}{5^2}\right] = 1$, $\left[\frac{31}{5^3}\right] = 0$, so 31! ends in 7 zeros.

Problem Set 4.4

1. $[360, 1970] = 70\,920$, $[30, 365] = 2190$.

2. $[220, 284] = 15\,620$, $[1184, 1210] = 716\,320$,
   $[2620, 2924] = 1\,915\,220$, $[5020, 5564] = 6\,982\,820$.

Problem Set 5.2

1. $m = 8$, $n = 1$, $(16, 63, 65)$, $n = 3$, $(24, 55, 73)$
   $\quad\quad\quad\quad n = 5$, $(80, 39, 89)$, $n = 7$, $(112, 15, 113)$
   $\quad m = 9$, $n = 2$, $(36, 77, 85)$, $n = 4$, $(64, 65, 97)$
   $\quad\quad\quad\quad n = 8$, $(144, 17, 145)$

$$m = 10, \quad n = 1, \quad (20, 99, 101), \quad n = 3, \quad (60, 91, 109)$$
$$n = 7, \quad (140, 51, 149), \quad n = 9, \quad (180, 19, 181)$$

2. No. If
$$2mn = 2m_1 n_1, \quad m^2 - n^2 = m_1^2 - n_1^2, \quad m^2 + n^2 = m_1^2 + n_1^2$$
it would follow that
$$m^2 = m_1^2, \quad n^2 = n_1^2 \quad \text{or} \quad m = m_1, \quad n = n_1.$$

3. When a number c is the hypotenuse of a Pythagorean triangle then every multiple $k \cdot c$ of c has the same property. Thus one need only list those values $c \leq 100$ for which no divisor of c is a hypotenuse. These one finds from the primitive solutions listed above:

$$c = 5, \quad 13, \quad 17, \quad 29, \quad 37, \quad 41, \quad 53, \quad 61, \quad 73, \quad 89, \quad 97.$$

Problem Set 5.3

1. $(120, 50, 130) \quad (624, 50, 626) \quad (48, 14, 50) \quad (40, 30, 50)$
$(120, 22, 122)$.

2. $100 = 10^2 + 0^2, \quad 101 = 10^2 + 1^2, \quad 104 = 10^2 + 2^2,$
$106 = 9^2 + 5^2, \quad 109 = 10^2 + 3^2.$

The numbers 101, 106, 109 are hypotenuses of primitive Pythagorean triangles.

3. There are no Pythagorean triangles with areas 78 or 1000. There is one triangle $(24, 10, 26)$ with area 120.

4. These numbers cannot be perimeters of any Pythagorean triangles.

Problem Set 6.2

1. 194 and 364.

2. $362 = (1, 0, 1, 1, 0, 1, 0, 1, 0)_2 = (1, 4, 0, 2)_6 = (1, 4, 5)_{17}$

$$1969 = (1, 1, 1, 1, 0, 1, 1, 0, 0, 0, 1)_2 = (2, 2, 0, 0, 2, 2, 1)_3$$
$$= (6, 13, 14)_{17}$$
$$10000 = (1, 0, 0, 1, 1, 1, 0, 0, 0, 1, 0, 0, 0, 0)_2$$
$$= (1, 1, 1, 2, 0, 1, 1, 0, 1)_3 = (2, 0, 10, 4)_{17}$$

Problem Set 6.3

1. The non-trivial multiplications are the products of a pair of the numbers 2, 3, $\cdots$, $b - 1$. Since the order of multiplication is immaterial we may take the smallest factor first. Then there are $b - 2$ products involving the factor 2, namely

$$2 \cdot 2, \quad 2 \cdot 3, \quad \cdots, \quad 2 \cdot (b - 1),$$

and $b - 3$ products with 3 as the smallest factor, namely

$$3 \cdot 3, \quad 3 \cdot 4, \quad \cdots, \quad 3(b - 1).$$

By continuing one sees that there are

$$(b - 2) + (b - 3) + \cdots + 3 + 2 + 1 = \tfrac{1}{2}(b - 1)(b - 2)$$

non-trivial products.

2. If one includes the factor 1 in the products one obtains them all by expanding

$$(1 + 2 + \cdots + (b - 1))(1 + 2 + \cdots + (b - 1)) = (\tfrac{1}{2}b(b - 1))^2.$$

For $b = 10$ this gives $45^2 = 2025$. If the factor 1 is excluded the sum is

$$(2 + \cdots + (b - 1))(2 + \cdots + (b - 1)) = (\tfrac{1}{2}(b + 1)(b - 2))^2.$$

For $b = 10$ one finds the sum $(44)^2 = 1936$. In both ways the sum is a square.

Problem Set 6.4

1. The function

$$f(b) = \frac{b}{\log b}$$

is positive in the interval $1 < b < \infty$ and $f(b) \to \infty$ when $b \to 1$ or $b \to \infty$. The derivative,

$$f'(b) = \frac{\log b - 1}{(\log b)^2},$$

vanishes only when

$$\log b = 1, \quad b = e = 2.71828 \cdots$$

and it is negative; hence $f(b)$ decreases in the interval $1 < b < e$, and increases when $e < b < \infty$. The minimal value is

$$f(e) = e = 2.71828 \cdots.$$

The function

$$g(b) = \frac{b - 1}{\log b}$$

is positive when $1 < b < \infty$ and

$$g(b) \to 1 \quad \text{as} \quad b \to 1; \qquad g(b) \to \infty \quad \text{as} \quad b \to \infty.$$

The derivative is

$$g'(b) = \frac{\log b + \dfrac{1}{b} - 1}{(\log b)^2}$$

and this is positive in the interval $1 < b < \infty$ so the function is increasing.

Problem Set 6.5

1. $2^n + 1 = (1, 0, 0, \cdots, 0, 1)_2$ with $n - 1$ zeros.

2. $2^p - 1 = (1, 1, \cdots, 1)_2$ with p ones, and so

$$2^{p-1}(2^p - 1) = (1, \cdots, 1, 0, 0, \cdots, 0)_2$$

with p ones and $p - 1$ zeros.

Problem Set 6.6

2. 92836
    12836
    ————
  105672

3. 29786
    850
    850
    ————
  31486

5. 411
  411
  411
  714
  ————
 1947

For problems 1. and 4. we leave you on your own. If you get desperate consult with some of your computer friends.

Problem Set 7.2

1. $-37 \equiv 5 \pmod 7$, $\quad -111 \equiv 10 \pmod{11}$,

$-365 \equiv 25 \pmod{30}$.

Problem Set 8.2

2. For $C = 19$ the last two terms in the formula (8.2.2) reduce to

$$[\tfrac{1}{4}c] - 2C \equiv 1 \pmod 7.$$

Problem Set 8.3

1. 1 : 2 : 3 : 4 : 5 : 6 : 7 : 8

———————————————————

7 : 6 : 5 : 8 : 3 : 2 : 1 : 4

———————————————————

8 : 7 : 6 : 5 : 4 : 3 : 2 : 1

———————————————————

2 : 1 : 7 : 6 : 8 : 4 : 3 : 5

———————————————————

3 : 8 : 1 : 7 : 6 : 5 : 4 . 2

———————————————————

4 : 3 : 2 : 1 : 7 : 8 : 5 : 6

———————————————————

5 : 4 : 8 : 2 : 1 : 7 : 6 : 3

———————————————————

6 : 5 : 4 : 3 : 2 : 1 : 8 : 7

———————————————————

2. When $r = 2$ the exceptional case occurs for $x = 1$; hence 1 plays 8 and 8 plays 1. For the other values $x = 2, 3, \cdots, 7$ one has

$$y \equiv 2 - x \equiv 9 - x \pmod 7$$

so correspondingly $y = 7, 6, \cdots, 2$.

3. Team $N - 1$ plays

$$y \equiv r - (N - 1) \equiv r \pmod{N-1}$$

in the r-th round. $N - 1$ can only be the exceptional team when

$$2(N - 1) \equiv r \pmod{N-1}$$

hence $r = N - 1$ and then $N - 1$ plays N.

4. The condition (8.3.2) is symmetric in x and y_r when x is regular. When x satisfies (8.3.3) it plays N and by definition N plays x.

References

We have presented you with an invitation to study number theory. If you are interested and wish to accept it you should continue by reading more advanced books on the level of college courses. There are many such books which can be recommended. I should like to mention first my own book:

Number Theory and Its History, McGraw-Hill Book Co., New York.

This book represents a natural next step since it deals in greater depth with some of the topics we have touched upon,and it expounds other theories of a more advanced nature.

There are also many other excellent books on number theory for college courses:

B. W. Jones, *The Theory of Numbers*, Rinehart, New York.

W. J. LeVeque, *Elementary Theory of Numbers*, Addison–Wesley' New York.

C. T. Long, *Elementary Introduction to Number Theory*, D. C. Heath, Boston.

N. H. McCoy, *The Theory of Numbers*, Macmillan, New York.

I. Niven and H. S. Zuckerman, *Introduction to the Theory of Numbers*, Wiley, New York.

H. Rademacher, *Lectures on Elementary Number Theory*, Blaisdell, New York.

J. M. Vinogradov, *Elements of Number Theory* (translated from Russian), Dover, New York.

The following books are somewhat more advanced:

H. Cohn, *A Second Course in Number Theory*, Wiley, New York.

E. Grosswald, *Topics from the Theory of Numbers*, Macmillan, New York.

G. H. Hardy and E. M. Wright, *An Introduction to the Theory of Numbers*, Clarendon, Oxford.

W. J. LeVeque, *Topics in Number Theory* (2 vols.), Addison-Wesley, New York.

D. Shanks, *Solved and Unsolved Problems in Number Theory*, Spartan Books, Washington.

If you wish to be informed about the history of number theory and find the authors of specific results you should consult:

L. E. Dickson, *History of the Theory of Numbers* (3 vols.), Carnegie Institution of Washington, Publication No. 256. Reprinted by G. E. Stechert, New York.

A general review of the tables of special numbers used in number theory can be found in:

D. H. Lehmer, *Guide to Tables in the Theory of Numbers*, Bulletin of the National Research Council, No. 105, 1941 and 1961.

INDEX